Outdoor Women inside the Forest Service, 1971–2018

Outdoor Women inside the Forest Service, 1971–2018

by
Lauren Turner

with a foreword by Abigail Kimbell

The McDonald & Woodward Publishing Company
Newark, Ohio

The McDonald & Woodward Publishing Company
Newark, Ohio 43055
www.mwpubco.com

Outdoor Women inside the Forest Service, 1971–2018

Text © 2018 by Lauren Turner

All rights reserved
This book has been printed in the United States of America by McNaughton & Gunn, Inc., Saline, Michigan, on paper that (1) meets the minimum requirements of permanence for printed library materials and (2) is free of harmful chemical contaminants.

Third printing August 2020

10 9 8 7 6 5 4 3
27 26 25 24 23 22 21 20

Library of Congress Cataloging-in-Publication Data

Names: Turner, Lauren Joyce, 1951- author.
Title: Outdoor women inside the Forest Service, 1971-2018 / by Lauren Turner.
Description: Newark, Ohio : The McDonald & Woodward Publishing Company, [2018] | Includes bibliographical references and index.
Identifiers: LCCN 2018032143 | ISBN 9781935778455 (perfect bound : alk. paper)
Subjects: LCSH: United States. Forest Service--Officials and employees--Women--History--20th century. | United States. Forest Service--Officials and employees--Women--History--21st century. | Women in forestry--United States--History--20th century. | Women in forestry--United States--History--21st century. | Women--Employment--United States.
Classification: LCC SD565 .T83 2018 | DDC 333.750973--dc23
LC record available at https://lccn.loc.gov/2018032143 (copyright page)

Reproduction or translation of any part of this work, except for short excerpts used in reviews, without the written permission of the copyright owner is unlawful. Requests for permission to reproduce parts of this work, or for additional information, should be addressed to the publisher.

Outdoor Women inside the Forest Service, 1971–2018

by
Lauren Turner

with a foreword by Abigail Kimbell

The McDonald & Woodward Publishing Company
Newark, Ohio

The McDonald & Woodward Publishing Company
Newark, Ohio 43055
www.mwpubco.com

Outdoor Women inside the Forest Service, 1971–2018

Text © 2018 by Lauren Turner

All rights reserved
This book has been printed in the United States of America by McNaughton & Gunn, Inc., Saline, Michigan, on paper that (1) meets the minimum requirements of permanence for printed library materials and (2) is free of harmful chemical contaminants.

Third printing August 2020

10 9 8 7 6 5 4 3
27 26 25 24 23 22 21 20

Library of Congress Cataloging-in-Publication Data

Names: Turner, Lauren Joyce, 1951- author.
Title: Outdoor women inside the Forest Service, 1971-2018 / by Lauren Turner.
Description: Newark, Ohio : The McDonald & Woodward Publishing Company, [2018] | Includes bibliographical references and index.
Identifiers: LCCN 2018032143 | ISBN 9781935778455 (perfect bound : alk. paper)
Subjects: LCSH: United States. Forest Service--Officials and employees--Women--History--20th century. | United States. Forest Service--Officials and employees--Women--History--21st century. | Women in forestry--United States--History--20th century. | Women in forestry--United States--History--21st century. | Women--Employment--United States.
Classification: LCC SD565 .T83 2018 | DDC 333.750973--dc23
LC record available at https://lccn.loc.gov/2018032143 (copyright page)

Reproduction or translation of any part of this work, except for short excerpts used in reviews, without the written permission of the copyright owner is unlawful. Requests for permission to reproduce parts of this work, or for additional information, should be addressed to the publisher.

Contents

Foreword vii
Acknowledgments xiii
Introduction. 3
Chapter 1: The Historical Background
   of the Forest Service 15
Chapter 2: Technicians 30
Chapter 3: District-Level Natural Resource Professionals . 72
Chapter 4: Forest-Level Natural Resource Professionals . 113
Chapter 5: Recreation Management Specialists 165
Chapter 6: Engineers. 194
Chapter 7: Unique in Some Way 213
Chapter 8: Line Officers 311
Chapter 9: Retrospective and Prospect 412
Endnotes. 437
Bibliography 441
Index 445

Foreword

A career in natural resources management is an amazing career. You work in places most Americans can only dream of visiting. You actively participate in sustaining life on earth. Conservation, or wise use of natural resources, is a calling. In the late 1800s and early 1900s, a handful of visionaries, including Gifford Pinchot and President Theodore Roosevelt responded to that calling, seeking to restore and preserve this nation's natural resources that were at risk from overuse. The United States Forest Service grew out of their efforts, with Gifford Pinchot as the first Chief of the agency.

Men were trained under Pinchot to venture out into the remote majestic mountains, grasslands and waters that blanket the country to police and protect those resources. Many were joined by women; wives, who worked steadfastly by their sides, with devotion, but without pay. History has recorded some of the wives stories, and eventually the inclusion of a few women employed in select positions in research or as fire lookouts. These women took pride in the connection of their work to caring for the land. As the agency and society evolved, women entered the Forest Service in greater numbers, and integrated into all the functions of the agency in this important work.

Outdoor Women inside the Forest Service, 1971–2018 is a collection of stories of women at the vanguard of the 60s, 70s and 80s, documenting their challenges and celebrating their accomplishments. Their stories are a pleasure to read. Each of these women is strong, independent and adventurous and more than a few are a tad rebellious. They came into a proud and hard working agency that didn't always embrace them, but they persisted and even thrived. They now stand as strong role models for young women today who may respond to the same calling. They worked in all aspects of the agency, and many went on to be supervisors, mentors and team leaders and have effected changes they had wished for, oh so long ago. These stories resonate with me. My first impulse after I read them was that I wanted to venture on a road trip to meet them all.

As you read these stories you will hear wonder, joy, surprise, disappointment and even anger. When these women faced adversity they stuck it out. They found ways to cope, found mentors and friends, in both women and men who worked beside them. They learned the value of teamwork. Their careers did not always turn out the way they might have envisioned, but it was so worth it. Reading their stories, you can follow their personal and professional growth, and trace the parallel growth and maturation of the agency.

Social change is never easy and yet it is constant. At the time I started my career in the early 1970s, the Vietnam War divided the nation. Returning servicemen and women were treated abominably. Title IX of the Civil Rights Act was signed into law. Environmental legislation was flying through Congress. President Nixon resigned office. Drugs and sex filled the news. The rules for living were changing rapidly. Women were coming into the workplace with freshly minted diplomas and ideas of equality. We just knew that if we worked hard, we would be rewarded with respect and a place on the team. We were eager to help make the world a better place for future generations.

Not all of our workmates saw this influx as a good thing. It wasn't an age thing or political difference — it was a struggle to deal with all the change. I recall an instance when one co-worker's wife went to speak with the boss and made it clear that if I were assigned to work with him, he would be forced to resign. Another fellow referred to me as his "dogger" and wanted me to bark in response to his commands. But I recall more instances where someone took me under wing and taught me everything I needed to know about a particular assignment. Each of us had to find or carve our way, often as the first or only. All of us had to mine inner strengths to be successful professionally and also as women. These profiles show how some of the women made it work for them.

Gender issues did not stop sometime in the seventies. It is part of having a mixed gender workplace. Even today, when I talk to young women in government or in private business, they will share that they do not always feel respected or valued. The workplace will always be fraught with interpersonal clutter. It is what we do with it that counts. Trying to write good behavior into hard policy guidelines has proven most difficult. There are equal employment opportunity laws and policies, zero tolerance policies, changing roles training and a host of other things. Still, leading by example and even-handed management are critical to any workplace. The lives chronicled herein include not only exceptional Forest Service employees; they reflect the evolution of an agency that has worked hard to embrace equality. The workplace has matured and become more contemporary, offering more opportunities to both women and men. These stories illustrate some of the tasks performed over the decades and demonstrate working through adversity, life changes and growth. Each path is very personal. Still, these footsteps illustrate so many opportunities for careers in natural resource management positions opened up to aspiring conservationists.

With opportunity comes choice, perhaps complicating decision making for individuals and for families. There are opportunities to pursue specialties in our professions. There are opportunities to work with other teams and on other units. There are professional societies offering after hours participation. Night courses are available on line. The communities we live in are nearly always very appreciative of Forest Service people serving as coaches, scout leaders, EMTs, planning commission members, hospital board members and such. And there are distasteful assignments that no one wants but that have to be done. Some choose to do all they can without leaving a geographic area. Others choose to see the world.

Perhaps the biggest choice is the one that partly defines us as women. Some choose to have children and I continue to be in awe of those that do. They share here what parenting throughout their careers has meant for them, some as single parents, some juggling dual careers. When I meet their children, these young people now, I love it that they don't think it odd their mother wears work boots and a hardhat or that she brings home live bats in a little bag or that she spends an evening presiding over a contentious public meeting. It is not odd that Dad may be the primary childcare provider in the family. In them I see hope that we really have made the social change only dreamt of forty years ago.

The environmental issues that challenge our nation's forests intensify with growing demands for resources. The Forest Service has studied climate change on forests and grasslands for more than thirty years and has some of the most comprehensive environmental data available. We are making tremendous efforts to deal with invasive species — plant and animal — but the complexities are most difficult to keep pace with. Despite our amazing water supplies, there are communities that lack safe drinking water and droughts are well publicized. Because nearly sixty percent of municipal water originates on forest lands, the Forest Service has a critical role in the health of the nation by maintaining the health of the forests.

These things may seem quite foreign to an increasingly urban population. But think about where wood comes from, where our water supplies originate, how we gather minerals for our technological tools, places to recreate and renew our souls — they come from the land that surrounds us. Whether we actively engage with it or not, we all need those resources from the land to be protected, to be managed wisely. More than ever we need people, like the women in this book, who will choose careers in natural resource conservation, who are excited when reading that such a career endows one with an understanding of "awesome" previously unimagined. We need those who can imagine pushing physically and mentally beyond their previously known boundaries, and want to know that personally.

I was privileged to know such experiences firsthand throughout my career. One of the real treats with retirement is having time to spend in the national forests, BLM public lands, wildlife refuges and other protected lands. One of my greatest joys is seeing a family on an outing, any kind of outing. Our world needs our children to grow into adults with at least an interest in natural resources if not a passion for their wise management. *Outdoor Women inside the Forest Service, 1971–2018* will inspire readers in that direction.

        Abigail (Gail) Kimbell
        Chief Emeritus, U.S. Forest Service

Acknowledgments

I first want to thank all of the women in this book that gave generously of their time to tell their stories, without whom there would be no book. Many of you expressed enthusiasm and offered me encouragement during the long journey it has been to get this into print, and I am deeply touched by that.

I am forever indebted to James Lewis, Historian of the Forest History Society, and author of *The Forest Service and The Greatest Good*, for his gracious support and guidance since the inception of this project. He shared his vast knowledge freely on several occasions, devoting his time to reviewing my work and advising me. He never indicated impatience, though I know his schedule is not easy.

Cheryl Oakes, Librarian for the Forest History Society was also notably responsive and helpful whenever I needed library services.

Forest Service Chief Historian, Lincoln Bramwell, PhD, was helpful on a number of occasions. I am particularly grateful to him for doggedly helping me track down supporting demographic data referenced in this book.

Sarah Baker, Wendy Campbell and Loretta Hall have writing and editing experience that they kindly let me draw upon for reviewing, developing and editing the book. Sarah introduced me electronically to Loretta, who does not even know me, yet Loretta gave me valuable, time-consuming advice and support on more than one occasion.

Bibi Gaston, great-grandniece of Gifford Pinchot, and author of *Gifford Pinchot and the First Foresters: The Untold Story of the Brave Men and Women Who Launched the American Conservation Movement*, came into my life during this process. I am in awe that she took the time to read and comment on this book, and offered her constructive advice and encouragement.

Thank you, Bibi!

My thanks, too, to Jerry McDonald of McDonald & Woodward Publishing Company for recognizing the value of this work, and for patiently guiding me through some of the technical aspects of getting it ready to publish. I am also grateful to the reviewers enlisted by McDonald & Woodward Publishing who offered support for my work as well as sound suggestions for edits that improved it.

My husband, John, has been a rock, which is reflected in my own story in the pages of this book. I want to thank him here for his patience, encouragement and good advice during this writing process.

Outdoor Women inside the Forest Service, 1971–2018

Introduction

Outdoor Women inside the Forest Service tells the stories of modern women who have worked in outdoor careers from 1971 through the present within the U.S. Forest Service (Forest Service), a historically patriarchal agency. This book adds a rich layer to not only Forest Service history but also to women's history. It chronicles the challenges and accomplishments of women in conservation, in the Forest Service in particular, that have never been told. It depicts a small slice of life during a pivotal time in history for both women and men.

In the 1970s, it was uncommon for the Forest Service to hire women for jobs other than clerical work. The civil rights movement of the 1960s, a combination of environmental laws and the women's movement of the 1960s and 1970s and a landmark 1973 class-action discrimination case which led to a consent decree as settlement in California culminated in opening new opportunities for women. Women entered jobs previously reserved for men in many different walks of life, in this case, the environment. By 2010, women represented thirty-eight percent of more than 30,000 Forest Service employees, and they held jobs covering every aspect of the Forest Service's business of "Caring for the Land and Serving People."

I worked for the Forest Service for thirty years, beginning in 1980, which was a career marked by opportunity and rich experiences. I worked in various clerical positions, and then mid-career I returned to college to qualify for professional field positions. I was a wildlife biologist, an ecosystem manager and ultimately I ended my career as a district ranger.

Reflecting back on my career after I retired, it gradually dawned on me that I had worked among a generation of extraordinary women who had contributed to a societal sea change, one of the most significant socioeconomic shifts in history. Many of them had experienced their most impressionable years during the 1960s, and began their careers in the 1970s and 1980s, times marked by changes around gender-based roles. I also came to realize that stories of these women's contributions to the work of the Forest Service during this period in history had not been told, and I decided I wanted to tell them.

My primary purpose in writing this book has been to document the contributions of these women in a personal, informal and informative way. These women have differed from many other modern women in that they have not dreamed of corporate boardrooms. They have headed to work in leather boots with Vibram® soles instead of stockings and four-inch heels. Their interview attire may have been jeans and a nice shirt. They have worked, and played, in the great outdoors — and many still do!

Cindy Champion's job, for example, is jumping from an airplane and parachuting to the ground deep in the woods to hike to the latest forest fire she will be fighting, wearing over a hundred pounds of gear (Figure 1). She and her crewmates set up

Figure 1. Cindy Champion suited up for a jump.

Introduction

Figure 2. Mary Westmoreland heading into wilderness to patrol for ten days.

camp where they will stay for one to several days. Mary Westmoreland will backpack alone for ten days in the Sierra wilderness, policing and educating the recreationists she encounters (Figure 2). Occasionally the visitors she meets will invite her to join their camp for a night, breaking up the sometimes-lonely nature of her solitary job. (Mary is a career Forest Service employee who declined to include her biography in this book, but she provided photographs, presented in the book, that are illustrative of her work in the outdoors.) Beth Humphrey loads horses, tack and gear into the four-horse trailer she will drive miles into the Sonoran Desert to spend a few days assessing the impact of grazing cattle on the southwest willow flycatcher, an endangered species of bird (Figure 3).

As my image of *Outdoor Women inside the Forest Service* matured, I grew curious to discover if these women whose stories I wanted to tell shared common characteristics that drew them into this choice of careers. I wanted to know about challenges they had faced in pursuing their careers and if they felt their

Figure 3. Beth Humphrey leading a pack string into Lime Creek Grazing Allotment.

presence in the Forest Service had changed the agency or their selves. In addition to their contributions to conservation management though, I knew that these women offered strong role models for other women, and that their stories might become beacons to attract younger women to jobs or careers in the Forest Service. Early pioneering women had paved the way for women entering the workforce during the 1970s and 1980s, and they in turn had played a role in improving the Forest Service work environment for younger women who came behind them. It is important to keep the trend going forward, for the Forest Service to recruit young women to keep pace where equality in the workplace is concerned. We need young women who are able to visualize their selves in this work that is so vitally important to sustaining our lives. By offering details about duties performed in the various Forest Service field jobs I hoped to spark their interest. With that goal in mind, I sought examples of women from a broad spectrum of Forest Service jobs.

In my effort to actually gather data, I reached out to women in the Forest Service I had known throughout my career who had chosen outdoor careers. Besides outdoor-oriented positions, women have made inroads into every field in the Forest Service, such as laboratory research, budget administration and human resources management. Though entering those

Introduction

jobs presented some of the same obstacles for women as those faced by women who chose outdoor careers, the stories of women in those fields are not the focus of this book. The social and especially the physical challenges were different for those in outdoor positions. I focused on the National Forest System because it is the branch of the Forest Service where most of the positions portrayed here are employed, but I included one example from each of the agency's other three branches — Research and Development, State and Private Forestry and International Programs, to further illustrate opportunities within other parts of the Forest Service.

I initially sent invitations to about eighty women who had worked in field-going positions for the Forest Service between 1971 and the present. I ended up with forty-one willing participants. I developed the following interview questions, which I sent to each respondent, and they answered by email or in telephone conversations. Answers to any of the questions were optional.

- Where are you from? City or country?
- Your background — activities as a child, likes and dislikes, favorite school subjects
- Family values as a child — did your family camp/hike/ski, or ...?
- In early grades, did you notice yourself being interested in/good at math and science?
- Did you feel more affinity with boys than girls? Did typical girls' activities bore you?
- In high school and/or college, did you notice being a minority in any of your classes? Which ones? What were the years?
- Some of you entered the Forest Service at a time when its field positions were nearly all male-dominated. Did you deliberately turn to the non-traditional?
- Have you faced any special challenges you'd like to share?
- Was there ever a time in your Forest Service career when you consciously felt that you didn't fit in/didn't belong? If so, what do you think evoked that feeling?

- Did you ever feel deliberately excluded or asked along as a token?
- Did you know any women who left the Forest Service because of male resentment or discrimination? Did you ever experience that kind of thing yourself?
- In either case, why do you think you chose to stay in the agency?
- What attracted you to your profession? What do you love most about it now?
- Please describe what you do — what does (or did) your day-to-day look like? (Enter here the chronology of your education and jobs you have held, including your current job. Please briefly describe each position.)
- What drives you/turns you on now? What is your passion? Your greatest goal in life?
- Who are the people in your life, what is your living arrangement?
- What do you do for recreation? Do you work out or is work enough exercise?
- What's your most extraordinary experience ever, or your top three?
- What did/do you see as challenges for women in the Forest Service?
- The Forest Service was changed by the entry of women into all areas of the workforce. Do you feel there was any way that working for the Forest Service changed you?

How did the Forest Service's reinvention of itself accommodate (or discourage, or both) the contributions of women? How has the work of women affected the shift in the Forest Service's identity and mission?

I compiled biographies based on these interviews that tell the stories of these women employed by the Forest Service during the past forty-seven years, most of who have lived through many of the major changes in the agency. These stories document their contributions to Forest Service natural resource conservation during this time.

These exceptional women told of their challenges and rewards and spoke to what drove them to want to do this kind of work. They described the kind of work that they did, informing us about the particular skills that are required in their professions. Notable in each of their stories is that they spoke from the heart. That they were women was incidental; that they found meaningful work was foremost. Some women sought jobs that were considered the most "manly," not because they wanted to be men, but because they wanted to do the work, and did not believe their gender should limit them.

There was a sense among the women of belonging to a culture of "Forest Service as family;" they tended to mentor and take care of one another. Most had a high regard for physical fitness. A common theme was that they could not believe they were lucky enough to be getting paid for work that seemed like fun or for being active in the outdoors amid spectacular scenery (figures 4-9).

Figure 4. Cheryl Hazlitt at Flaming Gorge Recreation Area — over the edge — work or play?

Figure 5. Joan Kluwe having fun with soil test pit in Bering Land Bridge National Monument.

Figure 6. Barb Stanley (right) taking a break from work on the front porch at Greytowers

Introduction

Figure 7. Mary Muchowski snowmobiling at Little Grass Valley Reservoir, Feather River Ranger District.

Figure 8. Mary Westmoreland getting around by snowmobile on Tahoe National Forest.

Figure 9. Traute Parrie (upper left) ice climbing in Beartooth Ranger District.

These stories illustrate a broad cross-section of the women in field jobs in the Forest Service. They offer a good representation of careers in the agency, since many have worked in more than one resource area and at more than one level. Some stayed in technical or lower-level professional jobs throughout their careers because they wanted to spend most of their time with their "boots on the ground" instead of balancing that with administrative duties that come with advancement. Those know their forests well, and their knowledge is vital to the agency's ability to carry out its mission. Others include luminaries, such as Abigail (Gail) Kimbell, first female chief of the Forest Service, and Anna Jones-Crabtree, an engineer who developed, and was the national director of the Sustainable Operations Collective, the first such position in the Forest Service.

The stories demonstrate the commonalities among these women, and the characteristics that distinguish them from other women. A common characteristic among these women is that they did not blindly accept society's expectations about their roles. Neither did they deny or discount their femininity. Some were rebellious, simply following their hearts when they chose their profession. I suspect that the somewhat stubborn nature shared by these women helped them persist in their careers when they encountered difficult times. While all have experienced the challenges of being a woman in what was once only a man's world, that is not the focus of their careers — it is, rather, the outdoors that they love and caring for the natural resources that sustain us all. They are remarkable not just because they are women, but because of their vision, commitment and determination to do the work.

This book speaks to a sense of adventure and a wide variety of "cool" careers filled with rich experiences. The stories illustrate the practical skills needed to work outdoors (driving on dirt roads, orienteering, map making, hiking and camping). I hope that they generate in readers a desire to go into one of these professions.

While the history documented in these stories is the focus of this book, I have included relevant information about Forest Service history, the agency's evolution, how the spectrum of jobs has broadened and how the roles of women in the agency have changed — from the unpaid working wives of male Forest Service employees to the wide variety of technicians and professionals of today.

Chapter 1 provides a brief overview of various federal agencies that manage and protect federal lands, it outlines the history of the Forest Service and its relationship to other agencies and it sketches out the evolution of women's presence and roles in the Forest Service. Chapters 2 through 6 illustrate the participation of women in different specialties and at different administrative or organizational levels within the Forest Service. Chapter 7 is devoted to positions or work that is unique in one or some ways, and Chapter 8 is devoted to positions of leadership within the Forest Service. For Chapter 9, I invited the women who I had interviewed to share in a frank discussion of Forest Service past, present and future outlook. To a woman, all acknowledged that the Forest Service, like any agency, has and will always have its issues, but all, even those who had faced profound difficulties, found that for them the positives outweighed the negatives and they were proud of and even grateful for having had a career with the Forest Service.

For those readers wanting further information, the Forest Service website offers extensive information about Forest Service careers. For specific position descriptions of various jobs and qualification requirements for those jobs, visit the Office of Personnel Management at https://www.opm.gov/policy-data-oversight/classification-qualifications/.

Chapter 1

The Historical Background of the Forest Service

Born of the Victorian Era, the Forest Service was understandably founded and run by men. For decades the agency remained staunchly patriarchal; nonetheless it has long attracted certain women to its ranks. These women have proven similar to their male counterparts in their love of the outdoors, their desire for adventure and their passion for caring for the land. Gradually as both society and the agency have changed, women have filled "non-traditional" field-going positions in the Forest Service, and today they are represented in every discipline at all levels of the agency.[1]

The Place of the Forest Service in Federal Land Management

The Forest Service is one of several government agencies entrusted with managing the public lands that all Americans own, in this case, 193 million acres of national forests and grasslands. The agency's mission is "to sustain the health, diversity and productivity of the Nation's forests and grasslands to meet the needs of present and future generations,"[2] which is summarized in its motto, "Caring for the Land and Serving People." Agency mission drives the types of activities that employees conduct on national forest lands. It determines the nature of the work of the positions it employs, and the nature of their interactions with the public.

Other prominent federal land management agencies include the Bureau of Land Management (BLM), the Bureau of Indian Affairs (BIA), the National Park Service (NPS) and the U.S. Fish and Wildlife Service (USFWS).

The BLM's mission, "to sustain the health, diversity and productivity of America's public lands for the use and enjoyment of present and future generations,"[3] is similar to that of the Forest Service. Both agencies practice multiple use of all the natural resources within their jurisdictions, providing goods needed by society, as well as conserved lands for watersheds, wildlife and recreation. The BLM manages 245 million acres of land, concentrated in the twelve western states, with an emphasis on energy management, most notably management of subsurface minerals. The Forest Service is recognized more for management of forests and grasslands and for fire fighting.

The BIA, whose mission is to "enhance the quality of life, to promote economic opportunity and to carry out the responsibility to protect and improve the trust assets of American Indians, Indian tribes and Alaska Natives,"[4] practices multiple-use management of natural resources similar to that of the Forest Service and BLM but across tribally held lands in the United States. Tribal lands are not public lands. They are held in trust by the federal government and are administered by the BIA. There is a government-to-government relationship between federally recognized Indian tribes and the federal government. The Forest Service is required by law to consult with recognized tribes on proposed actions that may affect tribal land.

The USFWS manages the system of national wildlife refuges — a national network of 150 million acres of lands and waters — with at least one refuge located in every state. Its mission is "working with others to conserve, protect and enhance fish, wildlife, plants and their habitats for the continuing benefit of the American people."[5] Both preservation and active manipulation of habitats are practiced on USFWS-administered lands. Hunting and fishing are extractive activities appropriate on lands in their jurisdiction. The USFWS also administers the Endangered Species Act, and so the Forest Service consults with this agency about activities on national forests and grasslands that may affect threatened or endangered species and their habitats.

The NPS mission is to "preserve unimpaired the natural and cultural resources and values of the national park system for the enjoyment, education and inspiration of this and future generations."[6] The public often does not realize the distinctions among these agency missions, thinking that they are one and the same, or not thinking about it at all. In the case of the Forest Service vs. NPS, the confusion is probably compounded by the fact that the two agencies' lands often abut each other, and you pass from one to the other seamlessly. The Stanislaus, Sierra and Inyo National Forests bound Yosemite National Park; Grand Canyon National Park shares its borders with the Kaibab National Forest; Yellowstone National Park borders the Custer Gallatin, Caribou-Targhee and Shoshone National Forests; the Olympic National Forest surrounds Olympic National Park. The Great Smoky Mountains National Park abuts the Nantahala, Cherokee and Pisgah National Forests. In some cases national park and national forest offices are co-located.

The Forest Service is the only principal federal land management agency administered under the U.S. Department of Agriculture; the others operate under the U.S. Department of the Interior. The main difference among these agencies lies in how they manage public land. The Park Service's mission is to preserve public lands, while the Forest Service and BLM missions are to use, or extract, natural resources wisely, to provide for the needs of the nation in a sustainable manner. The BIA mission is similar, providing for the resource and educational needs of American Indians and Alaska Natives. The USFWS mission falls somewhere in between, leaning a little more to the preservation side in many cases. All the agencies interact and cooperate. For instance, all are members of the National Interagency Fire Center, which was created to coordinate and reduce costs and duplication of efforts in fire fighting.[7] All agencies involved in federal land management also had responsibilities under President Obama's 2010 America's Great Outdoors Initiative[8] which called for federal, state and private entities to collaborate in connecting Americans to the outdoors, and in conserving and restoring America's great outdoors.

A History of the Forest Service and Its Mandates

The mission of the Forest Service is summarized in its motto, "Caring for the Land and Serving People." Caring for the land at first meant protecting and improving public forests, with priority on regulated timber production and security of stable water flows. Developing recreation areas and fighting fire received greater emphasis as part of employment programs established during the Great Depression, but timber production dominated Forest Service activities from its creation in 1905 through the 1970s. In 1960, in response to changing public opinion, Congress passed the Multiple-Use Sustained-Yield Act, giving management of fish and wildlife habitat, grazing, water, recreation and wilderness equal priority with timber in theory, though it would take several more years and laws before it did so in practice. Managing these additional resources requires further technical expertise, and the Forest Service today therefore employs a variety of natural resource managers, planners, engineers, researchers and geographical information specialists, among others. The agency has its own law enforcement organization. It has changed from an agency that strictly produced commodities to one concerned with ecosystem management and the sustainability of all resources, and from one that was managed internally to one that includes the public in its management. Beginning over 100 years ago as fifteen Forest Reserves staffed by a few employees, the Forest Service today has employed over 30,000 permanent employees and nearly as many temporary employees. The agency now employs over 200 clerical, technical, professional and administrative occupations. Most technical and professional employees work at least partially outdoors.

The Forest Service organization consists of four branches, each with separate governmental funding sources and distinct missions. The nine regions and six forest and range experiment stations under the National Forest System (NFS) are funded for the administration of NFS lands. Forest Research and Development, State and Private Forestry and International Forestry are the other branches of the Forest Service. This book focuses on jobs within the NFS, with an example from each of the other branches.

The National Forest System is organized into a headquarters — often referred to as the Washington Office or WO — located in Washington, D.C., and nine regional offices overseeing 154 national forests and twenty national grasslands (Map 1). The nine regions are numbered 1 through 6 and 8 through 10. Region 7 was absorbed into other regions some years ago. Each national forest and grassland is further divided into ranger districts. Ranger districts are the organizational level at which most of the on-the-ground work of the agency occurs.[9]

During the 1800s, before the establishment of the Forest Service, woodlands throughout much of the nation had been cleared to the point of shortage. Enter Gifford Pinchot, the first chief of the agency, who had acquired his parents' love of the forests and wanted to restore what had been destroyed. There were no forestry schools in the United States then, so he went to Europe to study. When he returned, he promoted sustained yield, which required intensive forest management, and he thought the land in the United States should be managed for all citizens.

"Caring for the Land and Serving People," meant to Pinchot that public lands should be managed to provide "the greatest good, for the greatest number, for the longest amount of time," which has long been the Forest Service mantra. The "greatest good" is a moving target, though, changing over time along with public opinion about how best to use (or not use) public lands. It is a balancing act figuring out how to responsibly exploit our environment in order to live, and to meet the needs of all the interests at the table. The Forest Service has a tradition of trying to do it all, priding itself on being a "Can Do" organization, but not every desired use is compatible on every acre of land.

In 1891 Congress passed the Forest Reserve Act, giving the president power to create forest reserves, by presidential proclamation, to be held by the people for the public good, specifically theft prevention, watershed conservation, fire protection

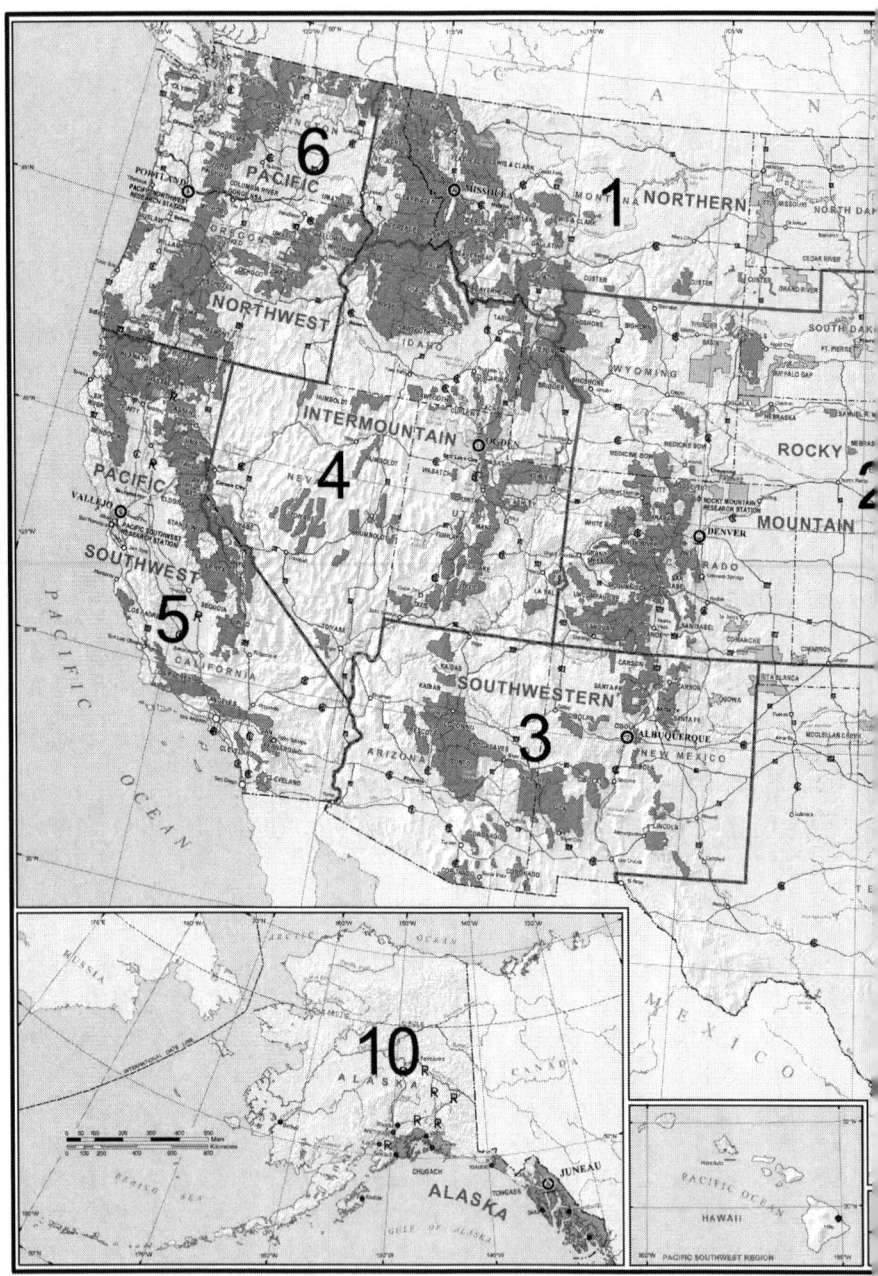

Map 1. Regions, forests and grasslands of the United States National Forest System, USDA Forest Service map. The names are (1) Northern Region, (2) Rocky Mountain Region, (3) South-

The Historical Background of the Forest Service

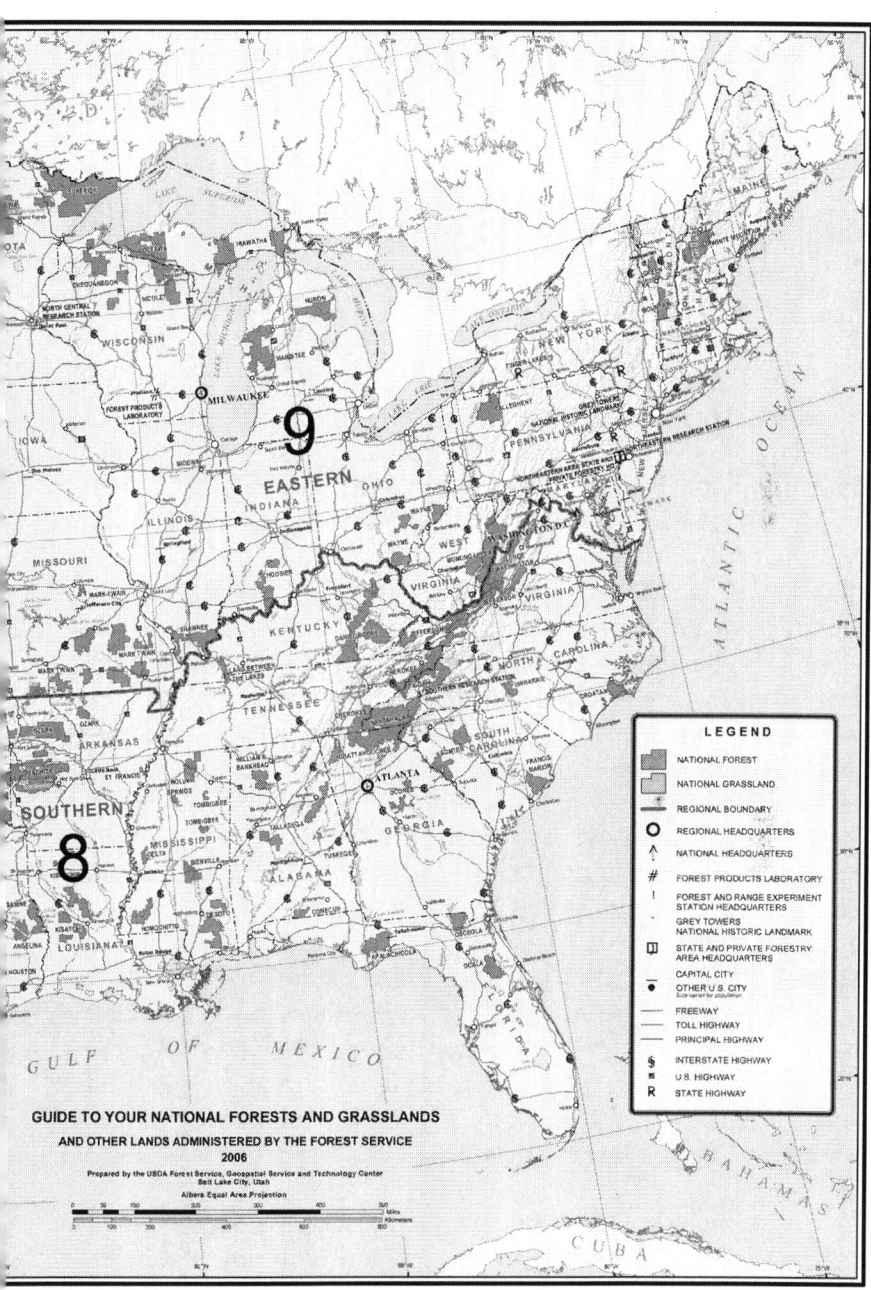

GUIDE TO YOUR NATIONAL FORESTS AND GRASSLANDS
AND OTHER LANDS ADMINISTERED BY THE FOREST SERVICE
2006

western Region, (4) Intermountain Region, (5) Pacific Southwest Region, (6) Pacific Northwest Region, (8) Southern Region, (9) Eastern Region and (10) Alaska Region.

and timber sale regulation. In 1898 President William McKinley appointed Gifford Pinchot head of the reserves. In 1905, under the leadership of President Theodore Roosevelt, a strong conservationist and Pinchot's friend, the forest reserves were transferred from the Department of the Interior to the Department of Agriculture, rechristened as national forests and placed under Pinchot's control as the first Chief of the Forest Service.

By 1905, a handful of forestry programs in universities had started around the United States, but only men attended forestry schools. There was a general forestry class included in Yale's summer field camp program, aimed at woodland owners, forest rangers and schoolteachers. Women were allowed to attend that general course, though they had to stay in town at a boarding house instead of at the camp with the men.[10]

Photos from that time show all-male forestry students, similarly dressed white men in suits, ties and horned-rimmed glasses, who would manage the national forest lands. A typical scene was of a man on a horse heading out to do forestry work in the field, his wife watching him leave. Those men would often be district rangers. In order to get the best men for the Forest Service, Pinchot had insisted upon civil service exams that tested for specific proficiencies.[11] The Forest Service developed the first written and practical exams for rangers by May 1906. Early recruitment posters stated: "invalids need not apply." Rangers were expected to "build trails, ride all day and night, pack, shoot and fight fire without losing (their heads)." Rangers were the lone stewards of several hundred thousand acres. As described by Robert J. Duhse: "The ranger in his district was often the only policeman, fish and game warden, coroner, disaster rescuer and doctor. He settled disputes between cattle and sheep men, organized and led fire fighting crews, built roads and trails, negotiated grazing and timber sales contracts, carried out reforestation and disease control projects and ran surveys."[12]

Early rangers were typically local hires respected in their communities. They lived on government compounds in remote areas, in tight-knit groups of a few other Forest Service families. Wives were expected to be sociable among the group and to manage things such as answering phones and radios and keeping things going at the district office while their husbands were gone to the field. Some wives accompanied their husbands, but they were not hired as field employees. There were essentially no women hired to do field work in the Forest Service for many decades. Wives were not paid for performing duties for the agency, but were considered disloyal and could harm their husbands' careers if they refused.[13] The experiences of the women who accompanied their Forest Service husbands to their remote posts is chronicled in a collection of their tales entitled *What Did We Get Ourselves Into?*[14]

The Forest Service has always employed a small number of women in clerical roles.[15] When it became clear that paper work was an important part of Forest Service work that the rugged rangers did not do very well, the agency hired women as clerks.[16] There still was some social rancor over women being in the workforce at all, and more rugged outdoor work was still deemed to be unsuitable for women.[17] A review of Forest Service history reveals a smattering of women working in various roles for the Forest Service, but their contributions are largely un-documented, names unmentioned. Photographs in *100 Years of Federal Forestry*[18], a comprehensive collection of historical photographs with text published in 1976 — coincidentally but tellingly, at the height of the women's movement — shows three women in skirts in 1927 transplanting tree seedlings, presumably at a Forest Service tree nursery, and a woman dispatching for fire in 1941. But there is no mention in the text of the women doing that work. There is a 1960 photograph of a group of Girl Scouts camping in the woods receiving instruction from a (male) forest ranger on proper use of map and compass for orienteering in wilderness. Interestingly, some of the women who tell their stories in *Outdoor Women inside the Forest Service* cite Girl Scouts as a childhood activity that was influential in their lives.

Gradually women gained prominence in the Forest Service and began to be marked by "firsts." These include the first female lookout, fire fighter, smokejumper and district ranger, on up through the first woman chief of the Forest Service in 2007.[19]

In 1910, huge forest fires broke out, killing people and burning entire towns. The devastation strengthened the concept that fire suppression is part of providing the "greatest good." The Forest Service built roads for better access to fires and lookouts in remote locations to detect fires. It was hard to find men to be fire fighters and lookouts. In 1913, before federal legislation allowed women to vote, Hallie Dagget applied for a lookout job in northern California. She was one of only three applicants, the other two were men: one was so disreputable that the fire ranger who received the applications refused to recommend him, and the other had poor eyesight and a gambling problem.[20] The Forest Service reluctantly hired Hallie, not expecting her to last long. She loved the position and stayed in it for fourteen years. During World War II, while the men were off to war, there was another surge of women hired as fire lookouts.

During the Great Depression in the 1930s, the meaning of "greatest good" shifted to using conservation projects to provide jobs and economic security. President Franklin Roosevelt created the Civilian Conservation Corps (CCC) to work on conservation projects on public lands, including national forests.[21] The conservation army of men planted billions of trees, fought fires, cleared trails, constructed roads and bridges, erected Forest Service office buildings and lookouts, built campgrounds and improved recreation facilities. The work of the CCC can still be seen all over the nation, and some sites have become monuments or have been preserved as visitor centers for the public.

CCC enrollees were young men over eighteen who were unemployed and agreed to serve for six months to two years.[22] Women were never included in the CCC, but a smaller, parallel program, dubbed the "SheSheShe" camps, promoted by Eleanor Roosevelt, provided job training to several thousand homeless women.[23]

The CCC concept was revived in 1957 as the Student Conservation Association (SCA), a non-profit organization that partners with corporations, foundations and agencies including the Forest Service, and places college students as volunteers in national parks and forests. The SCA continues today, providing hands-on conservation experience to high school and college students across the country.[24]

Agents of Change

Until the late 1960s, Forest Service field employees were mostly of the same ethnicity and gender, white males, and they shared a similar mindset. During the civil rights movement of the 1960s and 1970s, equal employment opportunity (EEO) laws, most notably Title VII of the Civil Rights Act, led to programs in government agencies to recruit and retain women and other ethnic minorities. Progress has been steady but slow since then, and not without growing pains. In 1972, Gene Bernardi, a former Forest Service employee in California (Forest Service Region Five), filed a discrimination complaint, Bernardi v. Butz, claiming she had been denied advancement because of her gender and that this discrimination was common practice within the Forest Service. She was convinced from the beginning that sex and race discrimination were systemic and widespread, and she wanted any changes resulting from her lawsuit to benefit all female and minority employees. In 1977, Bernardi requested the case be certified as a class action to include all permanent female general schedule (GS — the common pay plan in the federal government) employees working in Region Five as of November 5, 1972. The class action was granted, and the case was settled out of court in 1979, by a "consent decree," a legal agreement between the parties in which neither admits guilt, and that sets out terms negotiated by both sides for correcting the situation. The parties signed the consent decree in 1981.

The decree stipulated that Region Five's staff must be in line with the civilian labor force, with women in more than 43 percent of jobs in each series and pay grade, over the course of five years, as well as increasing the number of women at GS-11 through GS-13 levels for experience and exposure to training for higher administrative positions.[25] The court gave the Forest Service five years to create and implement a plan to meet the terms of the decree. The Forest Service did not meet that deadline, the Secretary of Agriculture was cited for contempt of court and the consent decree was extended for another five years. In order to meet the terms of the consent decree, the Forest Service ultimately offered training programs and formal education opportunities to help women at all levels to qualify for field positions and promotions in areas determined to be under-represented by women. The result was an influx of women into technical positions, which were often entry-level positions into the field, and an acceleration of advancement of women into line positions. The consent decree concluded in 2006.

Meanwhile, the CCC had become a model for later federal youth and conservation programs such as the Job Corps (1965), the Youth Conservation Corps (1971) and the Young Adult Conservation Corps (1977), which have fostered interest in land management and conservation among women and minorities. As part of President Lyndon Johnson's War on Poverty, the Job Corps program, starting in 1965, gave young men from disadvantaged backgrounds (teenage dropouts and military draft rejectees) basic schooling, training in skills and valuable job experience. It is a residential program that continues today and now is open to young men and women ages sixteen to twenty-four that meet low-income criteria. The Job Corps operates 124 centers nationwide. Twenty-eight of these are known as Job Corps Civilian Conservation Centers (JCCCCs), which the Forest Service operates on public lands under an agreement with the U.S. Department of Labor. The JCCCCs are located across twenty-two national forests and grasslands in eighteen states. Its initial emphasis was on fire fighting, fuels reduction and urban forestry, but the program has expanded to include other forms of natural resources management, recreation management and community projects that benefit the general public.[26] Many JCCCC graduates obtain permanent jobs with the Forest Service and develop into top leaders in the agency. Indicative of evolving cultural changes, the Youth Conservation Corps (YCC), a summer conservation work-learn program for young men and women from all parts of the nation, began in 1974, followed by the Young Adult Conservation Corps (YACC) in 1977, which provided enrollees with year-round conservation related employment and education opportunities. YACC operated at both the state and federal level. These programs provided gainful employment while accomplishing conservation projects and instilling in youth an understanding and appreciation of the nation's natural resource heritage. Due to funding issues, the Forest Service discontinued the YACC program after 1981. They continued the YCC program at a reduced level and the program continues today. Also, by then, individual states had begun to fund the YCC program directly.[27]

A burgeoning population, environmental disasters such as burning rivers and contaminated drinking water and increasing demand for natural resources coincided with many social movements during the 1960s and 1970s, including the civil rights movement, the anti-war movement, the women's rights movement and the environmental movement. There was a shift in public opinion from supporting natural resource extraction to a desire for the preservation of resources. This environmental movement resulted in new laws such as the landmark Multiple-Use Sustained-Yield Act of 1960, which shifted the Forest Service's emphasis from timber management by mandating management for a wide range of priorities — outdoor recreation, range, timber, watershed, forage, wildlife and fish — to best serve human needs. At the time the law was passed, ninety percent of professional positions were foresters. The law required balancing the competition for various resources so that the forests would be used conservatively and capable of supporting many uses with equal priority, in perpetuity.

Other laws enacted during that era that further altered the agency's priorities were the Wilderness Act of 1964, the Endangered Species Act (ESA) of 1973 and the National Environmental Policy Act (NEPA) of 1970. Today the Forest Service manages thirty percent of the lands designated as wilderness in the

United States.[28] Section 7 of the ESA requires federal agencies to insure that any action authorized, funded or carried out by them is not likely to jeopardize the continued existence of listed threatened or endangered species or modify their critical habitat.[29] As a result the Forest Service hired wildlife biologists and botanists to monitor and plan for species protection.

NEPA required government agencies to consider the impacts of their actions on the environment, and it required public involvement in the decision process. Forest and grassland administrators needed new expertise such as land-use planners and specialists (nicknamed "ologists") to work with the foresters who had been the predominant employees in the agency. They staffed interdisciplinary planning teams, with wildlife and fisheries biologists, botanists, soil scientists, hydrologists, landscape architects, recreation managers and archeologists, creating many new employment opportunities for men and women alike.

Another program, aimed at fighting poverty, was Volunteers in Service to America (VISTA), authorized under the Economic Opportunity Act of 1964 and founded in 1965. Later President Bill Clinton established the AmeriCorps National Civilian Community Corps (NCCC), a residential program for young men and women, and VISTA was incorporated into the AmeriCorps program. In the AmeriCorps program teams of youth are assigned to projects, including environmental projects. VISTA volunteers serve in communities, sometimes in creative environmental projects, designed to help ease poverty. The Forest Service participates in both programs. Both programs include higher education benefits to participants who complete the program.[30]

The parallel changes in the scope of the Forest Service's management responsibilities and in women's expectations of equal treatment and opportunity in the workplace contributed to more women becoming employed by the agency. While there have long been a small number of women hired to do limited tasks in outdoor natural resource conservation jobs, over the past few decades women have become involved in all aspects of national forest management. Women were admitted into the dangerous duties of smokejumping, fire fighting and law enforcement in the late 1970s and early 1980s. Janet Arling was the agency's first law enforcement special agent in 1978. Societal changes during the 1970s and 1980s created opportunities for women as experts in every discipline and as line and staff officers.

Beginning her Forest Service career in 1970 as a landscape architect, trailblazer Wendy Herrett became the first female district ranger in 1979. She became a forest supervisor in 1990 and regional director of recreation in 1992. Geri Larson was the first female forest supervisor in Forest Service history in 1985. Elizabeth Estill was the first regional forester in 1992, before and after which she held numerous other program director and deputy chief positions in the Washington Office. Estill was interviewed for the position of chief of the Forest Service position in 1996, but it was over ten years later before the Forest Service named its first female chief. More than a century after Pinchot's appointment, Abigail Kimbell became the first female chief of the Forest Service in 2007, serving through 2009.

Outdoor Women inside the Forest Service tells the stories of women employed by the Forest Service in field positions over the course of the past five decades, most of who have lived through many of the major changes in the agency. The following stories document the contributions of women to Forest Service natural resource conservation during this crucial time in history.

Chapter 2
Technicians

Technicians in various disciplines, such as biological science, hydrology, engineering and forestry, do much of the work that is performed on the ground in the national forests. These are the employees who form the backbone of the Forest Service, providing technical support to professional employees. Professional positions are separated from technical positions according to the educational requirements needed to qualify for the jobs. Professional positions require a minimum of a bachelor's degree or a defined number of college credits combined with relevant job experience equivalent to a bachelor's degree. Requirements for technical positions may be met entirely with on-the-job experience. The pay range for technical positions tops out at a lower level than for professional. The civil rights movement, California consent decree and environmental movement of the 1960s and 1970s, that expanded opportunities for professional women to enter the workforce likewise opened opportunities for the technicians who support them.

Technical jobs are often entry-level positions, providing a "foot in the door." Due to intense competition for Forest Service jobs, many technicians have worked seasonally for years before getting a permanent job.

Competition is compounded for women, as military veterans have preference for hiring, and fewer women than men are veterans. Some women have gotten around that barrier by first taking a clerical job to get on permanently. It may not be their first job choice, but women have traditionally held clerical jobs and there has been less competition with veterans for those positions. This is another irony of Forest Service history; men used to occupy even the clerical positions in the agency.[31] Once women were accepted as being suited to that work, those jobs became less valued, lower paying, dead-end jobs, with no easy access to higher-graded jobs, or even to lower-graded field positions.

The Land Management Workforce Flexibility Act, enacted on August 7, 2015, eased the long-term seasonal situation for technicians and other long-term seasonal employees. The act allows people who have worked as temporaries for the Forest Service and other federal land management agencies for a total of more than 24 months (does not have to be continuous) without a break in service of longer than two years to apply for jobs under internal merit promotion procedures, previously only open to permanent employees. This opens up a lot of job opportunities, and also puts all applicants on an even playing field, since veteran's preference does not apply to merit promotion. It does apply to the temporary jobs these employees originally compete for.

Frequently technicians are employees who have worked for many years, often their entire career, in one place, and they know their forest well. The Forest Service has a long history of requiring mobility to various posts throughout the country in order to receive promotions. While technical positions can eventually lead to advancement, they are also commonly held by employees who have resisted the pressure to relocate and to take on more administrative responsibilities in order to advance, prioritizing family ties and preferring to stay working in the woods, "boots on the ground."

The Forest Service needs technician support in a wide range of resource management activities, including fire fighting, improving roads, building and maintaining trails and planning and administering timber sales. This category also includes law enforcement officers, criminal investigators and special agents.

Technicians may serve as fire dispatchers or snow rangers, administer permits for special uses or conduct fire prevention and community education programs. Many conduct wildlife surveys or collect stream data or information about forest composition (tree size, species, age and condition). They collect vital information about resources that professionals use in environmental impact studies and for planning a variety of programs. Programs may include timber sales, stream restoration, fuels reduction, engineering, recreation, protection of archaeological sites, or wildlife habitat improvement.

Technical positions are also traditionally considered the most masculine of all the jobs in the agency, performed in often-adverse conditions, under extreme physical demands. Women not only had a hard time breaking into technical jobs as late as the 1970s and 1980s, but when they did they faced hostility from the men they worked with, and from those men's wives and even other female employees in the agency. In her book *Woods Working Women: Sexual Integration in the U.S. Forest Service*, Elaine Pitt Enarson summarized interviews she conducted with Forest Service technicians who worked in the field on two Oregon ranger districts. Enarson described the resentment toward female intrusion into technical positions such as trail, timber, or fire crews where the workers were together, even overnight, in the woods. She discussed the various coping mechanisms women used to deal with the hostility and discrimination directed at them, including violence in some cases, which varied from keeping quiet about it, just going along, joking and working twice as hard as the men to get the same recognition for their work. Some of the women I interviewed also spoke of using the same strategies and of having had the sense that they needed to work harder to prove they could do the work.

The biographies of the technicians in this chapter illustrate a variety of services that technicians provide, and the challenges and rewards experienced by these women during the course of their Forest Service careers.

∽

Carmie Biaggi, Assistant Dispatch Center Manager

As a child, Carmie Biaggi (Figure 10) recalls playing outside with friends, swimming all summer in the creek that runs through their small town in Paskenta, California. Carmie's favorite pastime was horseback riding.

Paskenta is a small rural town located about three hours north of San Francisco and Sacramento and just east of the 913,000-acre Mendocino National Forest. It was a logging and ranching community, with a small Forest Service presence, and home to a few sawmills, the largest of which was Crane Mills. Though their mill has closed since Carmie was young, Crane Mills still conducts logging from their property, and hauls logs with their log truck. The town's population is now around 100 — about half of what it was when Carmie was growing up in the late 1970s and 1980s.

Carmie's family fished in the creeks or at her paternal grandmother's pond, and hunted deer on her ranch. They ate everything they shot — "well, maybe not the ground squirrels," Carmie said with a smile. Her maternal grandparents spent their summers in the mountains, and Carmie's family visited them off and on during summer.

Figure 10. Carmie on fire duty.

There was always work to be done. The family fed and cared for animals, cut wood for their use and anyone else who needed it and helped at community events. Carmie's mom always made sure they were having fun while they were working, saying, "Why not have fun, but get the job done." Carmie brought that philosophy to her career.

Carmie's mom worked for the local logging company, piling brush. Carmie sometimes worked with her in summer. She told herself that she would never pile brush for a living. As she piled brush during a fire later on, she thought, "Never say never."

Carmie rode horseback with her mom in the wintertime to help her with a job caring for another rancher's cattle. For several years they also helped some friends who had a summer range permit in the Mendocino National Forest. Carmie thinks that is when she decided how cool it would be to have a job that paid to ride the range in the forest.

Carmie is one of the forestry technicians who found ways to have a satisfying career with the Forest Service without moving from her home and family.

In 1988, when she was eighteen years old, she started with the Forest Service on the Mendocino National Forest, in silviculture, because no range jobs were available. She was offered a cooperative education position, through which she could earn an associate's degree and then convert to a permanent position. After she received the degree, she was converted to a permanent position in silviculture. She loved the job and people she worked with, and happily worked outside in the mountains, hiking all day.

Her crew did a lot of work with other departments like timber cruising, resources and recreation, doing different jobs all the time. Carmie was trained in fire fighting as her Forest Service militia duties, which are collateral duties for supporting fire suppression when extra forces are needed. That experience eventually led to where she is in her career now. She still remembers her first big fire paycheck and being able to buy a pair of Whites smokejumper boots, an "awesome" experience.

In the early 1990s, timber positions were downsized because of new restrictions on timber cutting, and Carmie changed jobs after five years in silviculture. She could go to one of the fire engines as a lead fire fighter or apply for jobs on other forests. She liked fighting fire and did not want to move from her family and friends, so she was "off to another adventure in the Forest Service." But her heart was still in silviculture. She worked on her qualifications in fire on the engine while still assisting in silviculture, earning her qualifications as crew boss and engine boss in 1994.

During that time, the Forest Service was operating under the terms of the *Bernardi v. Butz* consent decree. She says she never thought of it as an advantage, but more of a challenge. She knew she had to work hard, have good work ethics, respect and earn respect from her coworkers and earn her qualifications for every position. She had great support from her supervisors and coworkers throughout her career. She knew there were some nonsupporters along the way, but she always "took their negative to make my positive better." She thought sometimes she had to step back to consider whether the negatives were valid and how she could improve or overcome them.

Carmie considers her first supervisor, Jim Henderson, to have been her best supervisor. He helped her develop into a supervisor and work leader. He showed employees how to be responsible for their actions and how to support their team, no matter what. They learned to stand behind each other or to take a "butt chewing" together and to deal with issues right then and there before the mole hills became mountains, or to take the credit or recognition they or others worked hard for.

Carmie became assistant fire captain on an engine, but still "detailed" back to silviculture. (In Forest Service lingo, a detail is a temporary assignment.) When she got married in 1999, she still did not have her desired range management job, but she and her husband had a ranch with cows and horses, so she felt she had the best of both worlds.

In 2000, she detailed to the Redding hotshots. It was a busy fire season, especially out of state. After that summer she decided she did not want to travel as much. The crew life was not for her anymore, and she wanted to have a baby. In 2001, she took a position in fire prevention (patrol), which allowed her to be home every night. Soon her family consisted of husband Matt, stepdaughter Sara, stepson Zach and new baby Nick (Figure 11).

In 2004, she decided to move into fire and law enforcement dispatch. It was different being inside all the time, but she spends her days off busy outside on her ranch. Her background in fire and all her other great Forest Service experiences helped her with this new adventure. Today she is the assistant center manager for the Mendocino National Forest dispatch. Besides feeling appreciative for the great people she works with and has known from being in the Forest Service for over twenty-five years, she still feels like she is being challenged by all the new computer programs and the ever-changing process of how the Forest Service does business. Dispatch gives her a

Figure 11. Carmie (second from left) and family on vacation in Hawaii.

sense of gratification, being able to interact with and be a service to the field and the public.

Because of having changed to fire service, which has an earlier retirement option, Carmie was eligible for retirement in 2017, at age forty-seven. She said, "I won't know until the time comes, but I expect to be starting a new adventure — well not new, but full-time on the ranch with my family."

~

Susan Cueva, Hydrology Technician

Susan (Figure 12) proudly hails from a fifth-generation California family that has lived on the same property in the Sierra Nevada foothills since the 1860s. Some of her ancestors traveled there on the historic Beckwourth Trail. They worked at gold mining, timber harvesting and cattle ranching.

Her great-great grandfather grew an orchard on the property and traveled to nearby mining towns by horse-drawn wagon, selling fruit. The family also had a sawmill and a quartz stamp mill on the property. Agriculture, forestry and mining were part of Susan's childhood. She said her background in these areas gave her advantages when she went to work for the Forest Service.

Susan's family home and property were one of very few in the area, and miles from town. She grew up in what some would consider an idyllic setting, in a mixed forest of ponderosa pine with oak, madrone, cedar and shrubs.

She was the middle child, with two older sisters and two younger brothers. They were the "ranch hands" and worked daily with their dad and mom. Most mornings (when not in school) Susan went with her dad to nearby ranches where their cattle were kept. There she would help feed and care for cows, haul hay and irrigate pasture. She learned to drive stick-shift trucks on the ranch. At home there were chickens to feed, wood to chop, cleaning chores (Susan said "ugh" to that) and cooking, which she liked. She disliked building barbed wire fence, and she did not like rules.

Susan spent a great deal of free time outside playing. She and her siblings created games out of what was available. She

Figure 12. Susan at a forest lake — part of her every day work environment.

played on rope swings, climbed trees, made bows and arrows out of willow and buckskin strings and, "most fun of all," rode her horse at full speed, often bareback, through her backyard of forest and meadows. She liked to roam and explore the forest alone with the family dogs. She liked everything about horses. She got her first horse of her own when she was eight years old.

The family took cattle drives many miles up to their grazing allotment on the Plumas National Forest in the spring. In the summer they would haul the horses to the mountains and round up the big steers to sell. Susan remembered, "I got to ride my horse at breakneck speed through the forest where, little did I know, I would eventually work." She learned to find her way around forest roads and learned the river and ridge names. When it was raining or snowing, she still had to go out and do chores, as "cows are a seven-days-a-week job." She liked adventure books, especially about horses, cowboys and Native Americans. Interestingly, she noticed early on that the women in the books did a lot of the chores and the men got to go out and have the fun on the horses.

Susan did not like sitting still for long. She found most schoolbooks dry and had to force herself to pay attention in grammar school. She got good grades, but found herself staring out the window, wishing she were outside. In high school, subjects that interested her were biology, history, English and physical education. Susan said, "I was the fastest girl in high school even though I am only five foot three inches tall."

Family values revolved around working the cattle ranch. They were active in the local agriculture community, including the Cattlemen's Association and the local Grange.

When the family went camping it was at the cow camp on their grazing allotment. Hiking was to round up cattle to bring them home before winter. The kids had to learn to walk quickly and take long strides to keep up with their dad as they roamed up and down canyons looking for strays or helped irrigate pasture in the valley. Susan's Aunt Ruth once told her, "We found our fun in our work."

Family-oriented values evolved from being a close-knit, multigenerational family living in one household and making a living on the same property. At dinnertime nine people sat around the same table, including Susan's grandparents.

Susan was first attracted to her profession, about which she said, "I did not know would become my profession," when she was eighteen years old. She wanted work near her family, wanting to find a way to earn a living working in a forest without leaving northern California. She called the La Porte Ranger District about the possibility of a secretarial position, "not thinking that any position in the outdoors would be available for women." She thought that at least she could look out the window at the forest. She did not know that the Forest Service had been ordered by the court to hire women in field positions, and she was amazed when she was told that the only position available to her was as a fire fighter. She told the office she would be right there, and drove to the ranger station. And so the adventure began. She started her career on the emotion, "Woohoo!"

When Susan first started working, it was challenging to work in a department that did not have a lot of experience employing women. She still has the powder-blue hard hat that she was required to wear so that she could be distinguished from the men! Some of the men thought that she could not do the work because of her gender and small size, but when they saw her wield her fire tools, and witnessed her driving skills, they had to reevaluate. Susan just saw the work that needed to be done and did it. Going to work for a government agency, she had to learn to navigate bureaucracy and take the proper channels to get the work done. She said, "That's an art form!"

After holding a number of technical positions over the years, Susan became a hydrology technician. What she still liked best about her job was outdoor work in the forest. She felt fortunate that she got to work in such a beautiful environment every day. She found it rewarding to help protect the forest and restore damaged areas (Figurer 13).

She also enjoyed the autonomy of choosing the projects she wanted to work on and being trusted to prioritize what was best for the forest. "All with my supervisor's permission, of course," she added. She enjoyed that her job allowed her to be active, and she met some wonderful lifelong friends in the Forest Service throughout her career.

Susan has earned college units from the University of the Pacific and Butte College, but no degree. One class that influenced her the most was "Outdoor Environments of California." Several naturalist books on the Sierra Nevada Mountains were required reading, including texts about trees, birds, mammals, flowers and rocks. They went to a different ecosystem each week to camp and study that environment. She took classes in geology and fire science. She liked classes that had an outdoor, hands-on, real-world component. In addition to college she has done on-the-job-training, as well as formal training

Technicians

Figure 13. Susan's Cedar Flat restoration project.

she attended while working various Forest Service jobs. She wanted to learn and be trained specifically for the skills she needed to do whatever job she had at the time. She also continued to learn independently through reading natural history and pioneer biographies and through her own curiosity about what she observes in nature.

As is common for many wanting to work for the Forest Service, Susan was a temporary seasonal employee for many years. She believes, though, that she would not have been hired into the male-dominated field as a fire fighter in 1974 if it had not been for the consent decree. She recalls that there were very few women in the fire camps.

The crew dug fire line with a variety of hand tools, carried hose packs and laid out hose along the fire line to get water from the engine to needed points along the fire line. They mopped up fires, which was very arduous and dirty work. Mopping up is done after a fire is controlled and consists of extinguishing all smoldering material along the fire's edge, ensuring logs and debris cannot roll across the fire line, making sure all burning fuel is burnt out, spread or buried to stop sparks from traveling, clearing both sides of the fire line of snags, rotten logs, stumps, singed brush and low-hanging tree limbs and searching for underground burning roots near the line.

Between fires Susan's crew received fire response training. They also performed station maintenance, including painting every building in the government compound.

That work was boring, compared to the adrenaline rush Susan experienced while fighting a blaze of fire. She welcomed the call to action when the fire bell rang.

The following season Susan wanted to try out another national forest, and commuted to a nearby district on the Tahoe National Forest. She worked on a ten-person brush disposal (BD) crew, initially with one other woman, who did not finish the season.

The crew created shaded fuel breaks (stacking sticks) using chainsaws and hand labor. A shaded fuel break is a 200- to 400-foot-wide area where fire fighters thin the tree stands; thinning opens up the forest crown to prevent the spread of fire. Crews also remove ladder fuels (low-hanging branches and brush) to prevent a surface fire from burning upwards into the crown.

Susan's crew was available for fire fighting anywhere in the United States. They traveled to many places in California, fighting fire in the backcountry. Susan was excited by being flown in small planes and helicopters over rugged mountains. She tells of one wildfire where she was way down in a steep canyon on the Klamath National Forest cutting line when a rock bounced down and hit her hard in the arm. She was trying to recover when her boss came by to see why she was slacking!

Susan's next opportunity came on the final day of the work year in 1976, while she was still on the BD crew. The crew had flown to Colorado to fight a late-season fire. While riding in the back of an army-type truck, she noticed some small cabins with barns and horses. The crew leader told her that the cabins were remote out stations of the Forest Service. Susan said she would like to work with horses, and he told her there might be an opening for a horseback backcountry patrol on the Truckee Ranger District.

That winter, Susan met with the supervisor of the recreation department at Truckee and applied for the job. Part of the application process was to go horseback riding with the supervisor on Forest Service horses. She said that many of those horses were a little crazy because many different people had handled them. Susan and another applicant demonstrated their horsemanship. The one who rode the best would land the job. Susan proudly announced, "That would be me" (Figure 14).

Figure 14. Susan in uniform, one of her favorite places — on horseback.

Outdoor Women inside the Forest Service

The district had purchased two young, untrained mules. Susan assumed mules would not be that different from horses. She quickly learned that mules are smarter than horses, they can kick to the side with their front foot and they can hold a grudge. Her muleskinner skills, "and colorful language" to go with them, developed quickly.

Susan patrolled forest trails on horseback, educating backcountry forest recreationists. This was the first time she'd been required to wear a Forest Service uniform. In those days, with women so new to field positions, the Forest Service had not yet designed uniforms that really fit women, and there were no women's sizes to order. The smallest men's size was way too big, so the district bought Susan two men's uniform shirts and sent her to a seamstress in Reno to have them altered.

Part of the job was installing trail signs and building hitching rails for horses. She removed trees across trails using both hand and power tools. She packed food and supplies with the mules for herself and the eight-person trail crew (Figure 15). She considered herself a good mule packer "when no eggs broke."

An unforgettable event was when she dug a backcountry toilet while her partner went off riding. She put all her energy

Figure 15. Susan leading her pack mule through the woods.

and frustration into that hole. At the end when she tried to climb out of the hole, she found she had dug so deep she could not get out! When her partner returned, he laughed. She thought it was funny too, after she got out.

She also remembers the chilling cold of the Truckee Ranger District, which is in the Tahoe National Forest. Sometimes Truckee can be one of the coldest regions in the country during the summer. The crew took no tents along, as tents were cumbersome and difficult to put up in those days. They slept under the stars with just a sleeping bag and tarp. Sometimes the ground would be frozen white in the morning.

Despite these kinds of experiences, or perhaps partly because of them, this was Susan's dream job. She was able to explore the high Sierra Nevada Mountains of California. She vividly remembers riding through the wildflowers and streams, even as the lower country baked in drought below.

The next summer funding dried up for the backcountry horseback patrol, and Susan moved on to a job on a three-person trail crew. They improved trail tread surface and cut logs and brush on miles of trails. They did a lot of hiking and camping out. Susan learned the proper way to get water off the trail, constructing dips and water bars, knowledge she used later in her hydrology job. She said, "I missed my horses, but the high country was still just as beautiful."

Meanwhile, Susan had met the man who would be her future husband. He was working on the Plumas National Forest in fire fighting, and she wanted to be closer to him. She visited the La Porte Ranger District, carrying with her a photo album of the work she had been doing on the Tahoe National Forest. She asked to talk with the District Ranger about a job and was surprised that she was able to get right in for a meeting. He looked at her photo album and said that he wanted to create a job for her as the wild river patrol. She asked what that would entail, as she was not a good swimmer. He explained that the Middle Fork of the Feather River had been designated a Wild and Scenic River in 1968, but that he did not know what trails were out there or what recreational use was happening. Susan was delighted to learn that the job would be on horseback.

From that experience Susan learned a valuable lesson about photographing her work. She has been taking pictures since and became the unofficial district photographer. She noted, "The saying that 'a picture is worth a thousand words' is so true."

Susan had one co-worker on the wild river patrol. The canyon and river were beautiful and remote. The two technicians camped out with horse and mule, working ten days on and four off. They located and improved historic trails into the Wild and Scenic River and made signs for the trails. Susan hauled mining and recreation trash out of the camps on the river, using her mule, Maxwell. The trails were too steep for her horse, Paiute, but Maxwell could navigate them like a mountain goat. Convincing Maxwell to leave Paiute at the top of the steep trails was a challenge, though. Susan muses that after working and hiking in and out of those canyons on a daily basis, nothing else ever seemed steep on her future jobs. She learned to avoid looking up when hiking out of those canyons after cutting brush all day because, "the top looked too far away."

Adventure and challenges abounded. The horses and mule would try to escape at night, so Susan often would not get much sleep. She made rope corrals and hobbled the animals, but sometimes she would still have to track them come morning. She would find them taking baby steps with their hobbles still on.

Susan also remembers encountering several large rattlesnakes on the trails. She and her partner had to wait until the snakes decided to move before continuing work.

She recalled one day when there was a large bear in her camp close to where she had tied her horse and mule. She quietly untied the animals and handed the horse to her partner, and then yelled at the bear. The horse and mule bolted, but Susan and her partner did not let go of their ropes, knowing if they did, they would either have to hike to find them or call for help on the radio. Forest Service employees judiciously avoid calling for help on the radio, as they do not want their mishaps broadcast. Maxwell the mule dragged Susan a fair distance while she kept yelling, "Whoa mule!" When she and her partner finally got the animals stopped and turned around, the bear was long gone.

Susan's next great adventure was a three-year break from work to become a mother. During her winter layoff, she had had a dream that she was pregnant. When she told her boyfriend, Rich, he said he had been thinking about having a child, too, and so it was time. The two were married on Coconut Island in Hilo, Hawaii. Susan stayed home with their daughter, Shannon, until the spring of 1984. The temporary jobs she had held offered no health insurance benefits; when their daughter arrived, Susan had to sell the horse and mule that she owned to pay the hospital bill. During her hiatus she enjoyed her baby girl, gardening, training horses and hosting basket-selling parties at peoples homes.

When Shannon was old enough for day care, Susan wanted a job that would allow her to be home at night with her family. She asked for a job at the La Porte Ranger District and was told she could volunteer in recreation. She said no thanks. The district called later, offering her a temporary job that had opened, taking care of campgrounds.

She called this "front"-country work, because for the first time in her career she had to work in developed campgrounds with recreational vehicles and many people who were not used to being in the woods. Duties included fee collections, cleaning bathrooms and educating and keeping peace among the campers in the "virtual small town that would be created around the lake every weekend." There were approximately 300 campsites and five campgrounds.

The first summer she drove a full-sized, stick-shift pickup with no power steering that she had to maneuver around the campgrounds and back into campground spurs. The first three days she just drove around and waved at the campers, afraid to talk to them. It did not take her long to adjust, though. She said of herself, "I have never been afraid of much and will speak up when I see something that needs attention." She said that she had to "quickly clean up my mule-skinner language and always be pleasant and professional." Susan says that this front-country job taught her the psychology skills she needed for dealing with all types of people. She experimented with different techniques every weekend and found that some worked better than others. People on vacation could be challenging. She said, "It's a lot like herding cows: don't spook them, and give them a way out."

Some seasons Susan did not have much help except for volunteer campground hosts, who could camp for free if they assisted with cleaning bathrooms and meeting and greeting campers. One time she came out from cleaning a bathroom holding a broom and was approached by a camper who told her that there was another ranger working there who was quite strict. The camper did not realize that he was talking about Susan, and she just agreed with him and said that he would do well to listen to that ranger. Her theory was that the cleaner she kept the bathrooms, the better behaved the campers were.

Susan was trained to do Level II law enforcement, which meant that she could write tickets but not carry a gun. She wrote very few tickets but did use her warning book. There were a lot of rules to enforce, but she concentrated on the big three: pay your fee, keep your dog on a leash and keep your vehicles on the pavement. She viewed her job as serving the people, and she wanted them to have a good time.

Finally, in 1988, after eleven seasons, Susan's recreation technician job became permanent. The job was classified as permanent seasonal, which still gave her winters off, but now with fringe benefits. She enjoyed her family time during her wintertime off.

So that she could work longer into the fall, Susan found assignments in other departments, such as timber and watershed. Eventually the district added a permanent position in watershed to assist the hydrologist and soil scientist. Susan had amassed a lot of work experience that qualified her for the job. She was hired as a hydrology technician in 1991, and she has retired from her career in that position. Working for eleven seasons as a temporary employee had been a challenge. She never thought that she would make it through thirty years to her retirement, but that benchmark did arrive.

In this job Susan monitored completed projects for soil and water protection, used GPS to map streams and flagged and tagged streams for protection during any future timber harvest. With all of the new technology since she started with the Forest Service, she had to learn many new skills, including computer software and GPS equipment. She chose to specialize in those kinds of equipment that would allow her to go to the forest instead of doing too much office work. She related one story from her pre-GPS days:

> Late in the fall I drove to the high county. As I got out of my truck it started to snow. I thought I would go out and work for a short time mapping streams because I had driven for an hour and a half to get there. When I turned back to my truck a blizzard had hit, and I could not find the truck. At that point I was wandering for a long time, looking for a landmark. I did not want to call on the radio unless I had to. Finally I saw some old sheep tanks at a meadow and knew where I was, which was far from my truck. I called on the radio, and my husband, the fire prevention technician, was walking by his truck and heard my voice and knew I was in trouble. He and another person who worked in timber came up and found me. By then there was six inches of snow on the ground. In this case my work ethic got me into trouble, and I have since been more careful. I wish I had had my GPS unit that day, but they did not exist then.

Susan found, developed and implemented restoration projects for degraded streams, meadows, roads and mining areas, which included gathering data on stream and soil conditions. She enjoyed taking a project from start to finish and then monitoring it in subsequent years. The camera has been her best tool. She said that you do not know how well you did your restoration until the following two or three years. If, in the end, she did not like the results, she did things differently the next time. She liked going back to the numerous locations she had helped restore and protect over the years. There are several meadows along roads that now are edged with large rocks, so no one can drive in them. She spent a few days using a McCleod to rake out ruts, before they put those rocks there.

Susan's versatility over the years has served the Forest Service well. She says that she decided early on in her career that you create your own reality. By showing up at the door and asking, she had several jobs created for her. Susan characterizes herself as a determined person who sets her goals and goes for it. She is quick to jump in and help where needed. She has often assisted in writing NEPA documents for various projects, surveyed for sensitive plants and written the botany report, surveyed for archaeological resources and written archeology reports, designed projects and been present at the project site when the heavy equipment operators were working, cleaning up with a shovel and spreading straw. She said that many of the skills she learned in her childhood have been useful at work.

Susan is the quintessential example of a true forest technician — that invaluable resource that forms the backbone of the Forest Service. She has put her family and love of working in the outdoors first and in doing so has demonstrated a characteristic that is shared among the women in this book: she has remained true to herself. Susan took opportunities when she saw them in order to have a successful career in the Forest Service. She tried to keep jobs that she felt she could do well rather than taking positions she felt she would not succeed in. She believes that staying in one job in one place for a long time gave her the continuity to lend her hand where needed. Really getting to know her district was beneficial to her work. Her situation was not without challenges, though. She reflected on her career decisions:

> I am deeply rooted in my home and community. I have really had no desire to move to different states or locations. Many people who move up and keep jobs in the Forest Service do so because they are willing to change locations. So

for me, rather than change locations, I changed my skill set to meet the needs of the positions that were available. When budgets shrank and we had to downsize, I would find my position scheduled to be gone. I had to either find a new position, or justify the position that I was already in and wait for funding to improve.

Susan was also a resource for new staff, educating them about the geography, past projects and history of her district. Her advice to young people, particularly women who have a love for the natural world, is that they should not be afraid to try new things. For their benefit she said, "There are many jobs within the Forest Service. If your first job isn't a perfect fit, try another! The more you know, the more well-rounded you will be and the more valuable your knowledge will be to the organization." Susan also proved valuable to the organization in being outspoken. She said, "I have been willing to share oppositional opinions [including with management] when I thought it was in the best interest of the forest and people who have been given the task of caring for our national forests. After all, the forests belong to everyone. We are just the stewards."

∼

Mary Muchowski, Biological Science Technician

Mary Muchowski (Figure 16) described herself as being from the city and the country. Born in San Jose, California, she lived in the suburb of Santa Clara until she was eleven years old, when her family moved to the tiny town of Philo, population about 400, and built their home on twenty acres. Mary has always had a split of city and country in her. She loves aspects of both and needs both in her life. She now lives in a small city and works in the forest, which balances the dichotomy for her. It seems also to have endowed her with a broad-based view of the world reflected in her career choices.

As a child, Mary enjoyed anything physical; games made up as kids usually involved some sort of outside activity. She always loved animals, and volunteered to take care of the class pets, such as rats, snakes and fish, taking them home

Figure 16. Mary Muchowski.

during vacations. She was enrolled in an alternative "Open Classroom," that her mother helped initiate, from third through fifth grade. The program required parental involvement, and students got to go on dozens of field trips each year with parents driving. Her favorite trips were to parks, a scientific marine wildlife research boat out of San Francisco and places like the Lick Observatory or the Exploratorium. She liked science, math and physical education in elementary and high school, and played on volleyball and basketball teams.

Mary's family camped often when they lived in Santa Clara, frequently going to the beach, the desert of Death Valley, Baja, California, or to a mountain lake. Her mother and a friend often took the kids to picnic in cemeteries when they lived in the city, which is probably why she said, "I see them as parks and still enjoy visiting them."

Mary and her siblings did downhill snow skiing and water skiing with friends. She did not hike much until she worked for the Forest Service. Now she enjoys hiking, which is fortunate, since the wildlife fieldwork she does requires a lot of it.

Her chosen career path seems a natural extension of the active, varied lifestyle she has always known. Mary's official title is biological science technician. She refers to herself as a field biologist. She has worked in wildlife on seven national forests, two BLM resource areas and one national wildlife refuge (NWR) in four different states.

She enjoys working outdoors where there is "fresh air, beautiful scenery, peace, no concrete, few if any other people, glimpses of wildlife (figures 17, 18 and 19) and adventure."

Adventure such as sliding down a hill with no way to stop yourself, stepping on a rattlesnake, getting poison oak, getting the truck stuck in snow or mud, losing the antenna on your radio and wondering if you will ever make it out of the sixty-acre brush field alive, or having legs so sore that you don't know if you can make it back to the truck.

Figure 17. Mary Muchowski's glimpses of wildlife—spotted owl 1.

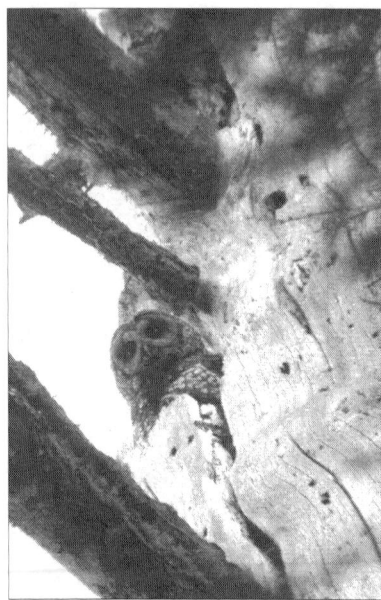 

Figures 18 and 19. Mary Muchowski's glimpses of wildlife — spotted owls 2 and 3.

During field season, Mary does wildlife surveys for California spotted owls, northern goshawks, bald eagles, willow flycatchers, peregrine falcons and amphibians (Figure 20).

Occasionally she surveys for forest carnivores (American marten and Pacific fisher), mountain beavers and bats. In recent years she has done a lot of map-making using ArcMap GIS software. She loves making maps. She fell in love with maps early in her career, during a summer in South Dakota when she saw her first topographic map and learned how to interpret it. She has been a "map-aholic" ever since.

Mary is another technician who has been a seasonal employee for years. The seasonal jobs are limited to a certain number of hours per year, usually during summer. Mary stretches her time to year-round work by working only twenty hours a week. During the fall and winter she writes reports, enters data and assists wildlife biologists by creating maps or analyzing data for upcoming projects or for NEPA documents. She prepares for the upcoming field season by designating calling points in suitable habitat for new project areas using GIS software, developing shape files which can then be transferred to GPS units and creating maps for use in the field.

Working for three different federal agencies, two under the Department of the Interior and one under the Department of Agriculture has given Mary perspective for comparing the three. She thinks that the Forest Service under the U.S. Department of Agriculture seems to be the most dysfunctional organization of the three, though the one in which she has worked the most time. She believes that some of this stems from the fact that the Forest Service is managed from the top down. She said, "And I mean the top! Congress decides how many million board feet of lumber we should be getting out of our forests." She thinks it would work much better if the local land managers on each district were able to manage the forest so that it will provide lumber and many other uses, such as recreation, and clean water from properly managed watersheds,

Figure 20. Mary surveying for goshawks.

for many years into the future. Many of the ranger districts that Mary has worked on no longer exist, as the Forest Service has combined several ranger districts into one during the past years of downsizing. Still, the agency continues to try to produce as much as ever.

Mary liked working for the U.S. Fish and Wildlife Service the most. Each national wildlife refuge is much smaller than a Forest Service ranger district. There are fewer employees, so everyone knows everyone. Each employee often gets to do a wider range of tasks. At Cibola refuge Mary was trained on the backhoe, farm tractor and skid loader. She helped mow fields, direct water to the fields, bush-hog brush along the roads and helped landscape around the office area. This was in addition to conducting wildlife surveys, planting native vegetation and running the Canada goose hunt.

Mary assesses that the BLM probably falls somewhere in between the Forest Service and the USFWS. Things seemed a little bit more relaxed at the BLM than at the Forest Service. When she worked for the BLM, they were participating in more research-based versus project focused wildlife surveys.

Her Forest Service days began in 1988, after she went on a student exchange to South Dakota State University, Brookings, from California State University (CSU), Chico, to check out their wildlife program. Before that, Mary says she did not know anything about the Forest Service when she "accidentally" got her first job with them that summer. Her ornithology professor had found a volunteer position for her, conducting bird surveys on the Custer National Forest in Camp Crook, South Dakota on the Sioux Ranger District. Her "pay" was supposed to be a stipend of $14 per day, but when she arrived, her boss had found a way to pay her and her co-worker a GS-3 wage.

Although Mary grew up in an area where logging was one of the main industries, there was not much public land around. There were a few state parks, but no national forests. She did not know that our national forests, parks or BLM lands hired wildlife biologists, and so had not really thought about it as a career. In college she majored in agriculture, animal science and pre-veterinary medicine. Once she went to South Dakota for the student exchange, took ornithology and got her first job with the Forest Service, she was hooked on wildlife and could barely muster any interest in her animal science classes. CSU, Chico did not have a wildlife program in those days.

Mary worked that first season on the Sioux Ranger District surveying for various bird species. She helped conduct stand exams in aspen stands, collecting age, canopy, tree/plant/herb species and wildlife use data. She entered the data on the then state-of-the-art Data General computer equipment (remember the days of the large floppy disc?).

In 1989, Mary was back in California, working on the Groveland Ranger District of the Stanislaus National Forest. She surveyed spotted owl habitat for fire damage from the Stanislaus Complex fire that had burned in 1988, and surveyed for spotted owls in proposed project areas. She took fire training, and worked on small lightning strike fires.

In 1990, after 6½ years at CSU Chico, Mary "graduated herself" from college. She worked for the BLM that season, on the Salem District of the Tillamook Resource Area in Tillamook, Oregon. She spent the next several years at various seasonal jobs in Oregon and California, bouncing back and forth between the BLM and the Forest Service. In 1996, she landed on the Feather River Ranger District of the Plumas National Forest, which was back in her old stomping grounds, near Chico.

Over those years Mary developed an array of technical skills. She has become expert at conducting surveys for California spotted owls, northern goshawks, amphibians, bats, bald eagles, willow flycatchers, peregrine falcons and forest carnivores. Planning and conducting wildlife surveys requires different methods, depending on the species. Besides setting up call routes for surveying for northern spotted owls, she actually handled the birds, learning to band adults and fledglings for later tracking. For carnivore surveys she learned to use sooted track plates and remote dual-sensor camera systems that helped positively identify animals that tripped the camera.

Mary performed other technical duties to support wildlife, fisheries, silviculture and fire. She conducted stand exams in plantations for silviculture, and installed rock weirs in streams for fish habitat. She planted riparian shrub species to improve riparian habitat. She worked with the range department, installing guzzlers (supplemental wildlife watering sources), building fences around sensitive riparian areas and maintaining stock ponds. She created maps, entered data, performed quality control checks of the database and wrote the inevitable reports. She fought fires as needed.

Mary left the Forest Service for a few years, working as an office manager and events coordinator for the Butte Environmental Council in Chico, and for several seasons for the USFWS on the Lower Colorado River National Wildlife Refuge (NWR) Complex in the Cibola National Wildlife Refuge in Cibola, Arizona.

For the Butte Environmental Council, she managed the office and volunteers, tracked finances and coordinated small and large events, including the annual Creek and Park Cleanups and the Endangered Species Faire.

On the NWR, she ran the Canada goose hunt program. She held a lottery each morning, and supervised the hunters to ensure they were following regulations. She surveyed two Breeding Bird Atlas blocks, a few point count routes, surveyed for clapper rails and other marsh birds and monitored shorebird use of flooded fields.

Since 1999, Mary has been working back on the Plumas National Forest, mostly on the Feather River Ranger District. From 2002 through 2007, she worked as a private contractor with the forest, splitting her time among its three ranger districts. The Forest Service had begun to contract most wildlife surveys, so she started a consulting business, Arroyo Chico Resources, with a Forest Service co-worker. Since their former jobs were being contracted out, they decided to be contractors rather than "contract inspectors" which they would have been had they remained Forest Service employees. As contractors, they got to continue doing the fieldwork themselves. As a contractor, she also tracked business finances and wrote proposals to bid on government contracts. In 2007, they discontinued their contract work. There were fewer contracts being awarded, and the stress of working for the Forest Service as a contractor had started to get to Mary.

She worked part-time again for five years for the Butte Environmental Council, developing programs and materials for educational outreach, financed by two large grants. She conducted research on the dioxin levels in free-range chicken eggs in the area as a follow-up study to an Environmental Health Investigations Branch study.

She has also been working part-time since 2007, back on the Feather River Ranger District, performing survey and data entry work. Her supervisor, the district wildlife biologist, considers Mary an invaluable asset to the district, and is grateful to have had her excellent skills to assist her over the years.

Mary's varied career has afforded her many extraordinary experiences. One that stands out is an evening driving to a spotted owl survey point, when she came to a large clearcut and startled a herd of elk. They scattered into the woods in all directions. At her call point she got out of her vehicle to wait for official sunset, when she could start calling for owls. The elk started calling to each other to regroup. Describing the moment Mary said, "All the calls together sounded strikingly similar to 'whale song'. It was an incredibly beautiful moment in nature, and one that I will never forget."

Another time, Mary was conducting a daytime follow-up on a pair of spotted owls she had heard the night before. She went down a watershed drainage while her co-worker went down another to the east of her. She found the pair of owls and started to settle in to watch them, to determine whether they were nesting. Suddenly she heard massive crashing noises coming towards her through the brush just east of her. It sounded like a herd of elk coming at her. She panicked, thinking, "What do I do if a herd of elk is going to stampede me?" She decided to stand super close to a large tree and wait. A bear came tumbling out of the brush field and into the open part of the forest, about fifty yards from her. She was standing perfectly still by the tree, amazed that one medium-sized bear could make so much noise running through the dense shrubs, and relieved that it was just a bear and not a herd of elk. The bear slowed down and stopped just after entering the forest. He ambled around for a little bit, and then walked directly under the pair of spotted owls that Mary was watching! "And me without my camera!" she lamented. She remained still, and the bear never noticed her as she watched him make his way down the hill. Later her co-worker said that she had startled the bear when going down her route.

Mary's career path has had its challenges. She has been a temporary employee for the Forest Service, BLM and U.S. Fish and Wildlife Service for over twenty-five years, partly by choice. She loves being a field biologist, and has noticed that if you move up, you become an "office biologist." Money is not much of a motivator for Mary, so she has stayed in the lower pay grade in order to remain a field-going biologist.

However, being a temporary has meant that she has not had any health or retirement benefits. For many years Mary worked summers on one temporary appointment, then worked winters in another location on another temporary job for a different agency. For several years she has worked part-time, year-round at one location for the Forest Service. This worked well while she had another part-time job at a local non-profit that offered health benefits. When she got laid off from the non-profit job, she lost her health insurance. Lack of health benefits and a retirement plan are probably the biggest downsides for her working as a temporary employee.

A position exists, which Mary said, "I have always wished, begged and hoped one of the agencies would put me in" that is called an eighteen and eight. It is a permanent seasonal category, in which you are guaranteed work for eighteen pay periods, and then are off for up to eight pay periods. It is still seasonal work, but with benefits. Mary thinks this would be the perfect situation for her, allowing her to still be able to do what she loves, be able to have health insurance and only having to be off for a maximum of eight pay periods.

When Mary first went to work for the Forest Service, she had not heard anything about the agency being "male-dominated," so she did not think about it being non-traditional for her to work there. "Not that that would have stopped me if I had known," she mused. However, after being at the ranger district for a couple of months, she heard how the other male permanent employees treated her supervisor, who was one of the first women resource managers to be hired there. She was not sure if the harassment was because she was a woman or because she was a biologist, or both. Her supervisor ended up taking another position in Idaho about half way through the field season. Mary does not recall being harassed in any way in South Dakota, or at any of her federal jobs. She speculates that may have been because she was only a temporary field biologist, and did not have to interact much with any of the other permanent employees, or anyone other than her field-going co-workers, who were also women.

In later years, she has known about other cases of harassment of women that worked on her district. One female fire fighter on a district filed an official complaint after being harassed for quite awhile. As a result, one male fire fighter was let go and several were given warnings. Another woman, a few years later, was harassed by several of the fire fighters on another district, although Mary does not think that an official complaint was made. More recently, one of her female co-workers said that a male Forest Service employee who she had to work with occasionally was always saying inappropriate things to her. She did not file a complaint, and did not want any of her co-workers to file a complaint on her behalf, or say anything. The Forest Service does require sexual harassment awareness training on a regular basis.

Mary's passion is bird watching, and her greatest goal in life is to share the natural world with others and to get them excited about it. She includes her niece and nephews in that. She proudly remarked, "They learned how to look through

Figure 21. Mary up close and personal with scenery in Painted Canyon, Cibola, Arizona.

binoculars at a very young age!" They have been participating in back yard bird counts with Mary since they were about four years old. Mary is currently the compiler for the Chico Christmas Bird Count for the National Audubon Society, and helps lead bird-watching trips for the annual Snow Goose Festival in Chico.

For Mary, a benefit of being a temporary for so long was being free to move around. She has "tromped around more federal lands than most people in this country, even most Forest Service employees, because I have been a field biologist the whole time. I have met lots of great people working for these agencies, and have seen many beautiful parts of this country, close up and personal (Figure 21)."

She said this was great when she was younger, but now that she is older, and wants to stay closer to her parents and family, and stay in one place. That also makes it harder to move up in a federal organization. If you are willing to move, you can often find a job to move up into. In Mary's case it is not the mobility expectation, but her desire to stay working in the woods that has kept her from advancing.

Another benefit of staying in the field, is not having the stress of being responsible for employees, and not having to deal with the forever changing Forest Service deadlines. Well, Mary admits she had to deal with the deadlines, by revising map after map when boundaries of the project area or the treatment of the units were changed, but she said, "I didn't have to take them seriously — no overtime for me."

~

Erin Vonderheit, Fire Dispatcher

Repeating a common theme running through the lives of the women in this book, Erin Vonderheit (Figure 22) was very active as a child. Her parents exposed her and her brothers to cross-country skiing at a young age, as well as camping, hiking and swimming. As Erin got older she became involved in soccer, T-ball, downhill skiing and running. She loved school and was an excellent student. She loved math and science and wanted to be a doctor when she grew up. She also loved to read and was very artistic.

Figure 22. Erin Vonderheit.

Erin was born in Dallas, Texas but was raised in Boise, Idaho, where her family moved when she was three.

Her family did a lot of camping, bike riding, hiking and swimming during the summer months and skiing and sledding in the winter. When she was in elementary school, around fourth grade, her parents bought a cabin in McCall, Idaho, so they spent most weekends at the lake during the summers and at Brundage Mountain ski area during the winters.

Erin graduated from high school in 2003. She attended Claremont McKenna College in California where she ran the lifeguard program, and played rugby and soccer and briefly ran cross-country. She was a resident assistant her senior year. During the summers she worked at a public pool in Boise as a manager, lifeguard and a swimming instructor.

In 2007, Erin graduated with a degree in molecular neuroscience and a sequence in women's studies. She was pre-med, but decided not to take the mCATs because she was not ready to commit to the debt, or the eight-year obligation of medical school.

From June through November of that year she worked her first season for the Forest Service in Bend, Oregon. She moved back to Boise after that, and taught eighth grade earth science for spring semester as a long-term substitute teacher.

Erin said, "I kind of fell into my job with the Forest Service." She had a friend who was going to fight fire in Bend for the summer who convinced Erin to join her. Erin called the fire crew supervisor, and the next thing she knew she was heading to Bend to fight fire for the summer, as a member of an engine crew. Growing up in Boise, she had been exposed to forest fires, but she had never considered joining a crew. Boise is home to the National Interagency Fire Center, which houses the National Interagency Coordination Center (NICC), responsible for coordination of all wildfire fighting resources in the United States, including those of the Forest Service. Even so, there was not a big push to educate children in the Boise schools about fire fighting careers. As a rookie seasonal employee on the Deschutes, Erin attended "Guard

Figure 23. Erin (left) early in her fire fighting days.

School" where she learned basic fire fighting, took the required classes to be a qualified fire fighter and had her first experience with a live fire drill. She discovered that she loved being on the line fighting fire and could not wait for the season to begin (Figure 23).

The excitement never left her, and Erin returned the next summer to work on a heavy engine for the Bend-Fort Rock Ranger District on the Deschutes National Forest.

Wanting more challenge than the on-district crew job, Erin looked for employment with a Type 1, rappel or hotshot, crew. In 2009, she worked on the Logan Interagency hotshot crew (IHC) in Utah (Figure 24).

After one season, she realized she wanted to be back in Oregon with her boyfriend, Trent, who she had met in 2007. She took a position on Winema IHC in 2010. Being on a Type 1 crew was an experience that she treasures. Type 1 crews

Figure 24. Erin (front, 2nd from left) Logan Interagency Hot Shots.

are viewed as a somewhat elite group of either hotshots or smokejumpers. Hotshot crews require a higher level of physical fitness than district crews, and Erin enjoyed pushing herself physically as well as getting into much more dynamic fire situations (Figure 25).

An extraordinary experience that really sticks out in Erin's mind was on her first fire with Logan hotshots in 2009. It had been a slow season. They had been to two floods before they ever went to a fire, and that fire assignment was not until July. They were finally ordered for a fire in the Book Cliffs of Southern Utah/Colorado. When they arrived at the fire after two days of driving, they realized that it was out on a knife ridge surrounded by cliffs. They spent the night on a plateau looking out at the fire, woke up the next morning and started their hike into the fire as the sun was rising. They would spike out near the fire, as it did not appear to be spreading rapidly, nor was it that large, according to reconnaissance that had been done the day before. Erin's crew was sent to the fire because resource advisors wanted the crew to protect an archeologically significant area of the canyon below the burning area.

After a few hours of hiking on some incredibly steep slopes, the crew finally reached some black. Unfortunately, the fire had slopped over the ridge and was much larger than anticipated. There was no way for the crew to take action on this new part of the fire due to terrain, and there was no place for them to sleep. As it got later in the day, and the fire activity increased, they decided to leave. They built a helispot by moving some very large boulders, and called in a helicopter to get the crew out. That afternoon Erin had the first helicopter ride of her career in a T3 helicopter in the Book Cliffs.

Erin was raised in a family that told her she could do anything. She is aware of her limitations, but has never thought of her gender as being one of them. Therefore, while fighting fire was not a "typically female" job, it never occurred to her

Figure 25. Erin cherished dynamic fire situations.

that she would be the minority when she took her first job on a district engine.

She thinks the fact that hotshot crews were traditionally male-dominated did affect her decision to join Type 1 crews. She welcomes challenges, and she felt that she would be a good female representative on those crews. She wanted to be held to a higher standard than the district crews, and she wanted to show that she could hold her own, physically, on a crew dominated by guys. Type 1 crews are required to have more training, overall, and are held to a higher physical standard than Type 2 and district crews. Most fire fighters start their careers on-district where the demands are somewhat less and they can "get their feet wet" before they decide to transition onto a Type 1 crew. Districts vary as to how busy they are, and the Bend-Fort Rock was a great place to start.

Erin liked the physical challenge, but was unprepared for the lack of emotional support provided by a primarily male crew. She made many great friends on the crews that she was on but a few "bad eggs" ultimately contributed to her decision to seek a job outside of fire operations. She said that she experienced discrimination because of her gender, and possibly because of her education and intelligence. The culture of a fire crew is to be manly and tough, and the small number of people that were either threatened by her or had no respect for her because of her gender made her last year on a fire crew miserable.

Those issues aside, Erin enjoyed the physicality of her fire fighter job, but started to miss being challenged mentally. After being in primary fire for four years, she was ready for a more intellectually stimulating job.

After four years in operational fire, two years on an engine and two on hotshot crews, Erin decided to move to dispatch. She was becoming a little jaded over the lack of opportunity for young, intelligent women on Type 1 fire crews. Though the job offered physical challenge and adventure, being a "grunt" on the fire line was too one-dimensional for her. She needed a job with multifaceted responsibilities. In 2011, Erin took a job at the local interagency dispatch center and did not look back.

She enjoys the intensity and dynamic nature of dispatching, and she relishes being closer to home and having a much more varied job than one does as a fire fighter.

She moved to Prineville with Trent in 2011, when he took a job there, and she took a job with the Oregon Department of Forestry (ODF) Interagency Dispatch Center as an initial attack (IA) dispatcher. She liked being back in Central Oregon where she had started her career with the Forest Service, and she worked with a great group of people, most of who had also been fire fighters during their careers.

The Forest Service offered better opportunities than the state of Oregon, so in 2012, Erin moved back to the Forest Service, at the same interagency office and still as an IA dispatcher. Working for the Department of Forestry in her inaugural season afforded the opportunity to continue dispatching for the state with the Department of Transportation in the fire off-season.

With the Forest Service, Erin had the opportunity to travel to other dispatch centers in the region, and to go to expanded dispatches to support large fires. Expanded dispatches are set up temporarily during large fires, at locations near the fire being managed. Employees who are specially trained for fire dispatch in addition to their normal jobs staff them. They are part of the Forest Service militia, who serve in short-term details to support fire as needed. Supporting extended attack is much different than initial attack. Depending on their availability, resources may be ordered from anywhere in the country, or even Canada, which is coordinated through NICC. Things like their location and time spent on their assignment must be tracked closely. Erin learned a great deal about the logistics that are required to support such large, complex operations.

As an initial attack dispatcher it is Erin's duty to track local resources and send them to incidents as they arise. Her day consists of tracking the location and availability of engines, crews and single resources (individuals such as safety officers or timekeepers) in a computer-aided dispatch system, taking reports from lookouts, aircraft, 911 dispatch centers and the public and dispatching resources over the radio and telephone. She is also tasked with all the documentation for these incidents. Permanent full-time employees have other responsibilities which include keeping up the remote automated weather stations for the Deschutes and Ochoco National Forests, keeping training records for all employees and coordinating classes for them, filing fire reports and preparing for the upcoming fire season. Dispatchers also act as liaisons during preparedness reviews and district fire meetings. They go out in the field to educate employees on what they do and how dispatch and ground personnel work together.

Erin has a passion for learning, and for being challenged. She loves being busy at work; she likes organizing things and being ready for the next big fire. She likes being part of a team. She works in an environment that is constantly changing and has great coworkers that appreciate suggestions on how things can be done better, or differently. She likes that she has the opportunity to tailor her job to herself and is given the freedom to go about tasks in a way that makes sense to her, as long as the end result is the one that was desired. Coming from a more-rigid fire organization, she savors the freedom.

Erin wants to be able to provide for herself and her future family. She would love to be able to contribute enough, financially, for Trent, now her husband, to continue to work in the field and not have to make career decisions solely based on his ability to support a family. As someone who was raised in a stable "all-American" family with a hard working father and a stay-at-home mom, she sees great benefit in being able to stay at home with her children. Contemplating motherhood, she struggles with the idea of having to put children in daycare. At the same time, she enjoys working and would be extremely happy to provide for her family and make her husband's life easier by sharing the responsibility of supporting their family financially.

She and Trent have discussed how to both keep the careers they love with the Forest Service and to start a family. They feel very lucky to have co-workers who are great examples of dual-working Forest Service parents with great families. Erin said, "If it were not for the women who came before me, forging the way in a 'man's world' and doing 'man's work,' as well as the men who accepted them and love them, I would not have these great examples to look up to."

One of Erin's greatest goals is to excel at her job and to be a positive force in the workplace. She greatly enjoys dispatching, but does not see herself in this vein of the Forest Service forever. She does, however, see herself working in a land steward agency. Although it is much more difficult for a white female, with no veterans preference, to obtain a permanent job with the federal government these days, the benefits of staying in the same, or similar, field as her husband, as well as the enjoyment she gets from her job and her coworkers, outweighs the hardships of being a long-term seasonal employee. She has considered getting a master's degree in a field that would increase her chances of being hired as a full time employee, but there are not many options for higher education in the area she lives, and she does enjoy what she is doing for the time being. No doubt, she will get it figured out. With her energy, enthusiasm and intelligence, whatever path she chooses, she will be sought after.

Chapter 3
District-Level Natural Resource Professionals

The ranger district is often the public's first point of contact with the Forest Service. There are more than 600 ranger districts in the Forest Service, ranging in size from 50,000 acres to more than a million acres, each employing ten to 100 people.[32]

Women in various natural resource areas on ranger districts are profiled in this chapter, including archaeologists, silviculturists and wildlife biologists. The Forest Service also employees a broad range of other resource professionals, including botanists, fisheries biologists, engineers, soil scientists, hydrologists, entomologists, recreation specialists, land management planners and others. All district-level employees work under the leadership of district rangers. Profiles of district rangers are included in *Chapter 8, Line Officers*.

This chapter includes the first story in this book of a woman who spent part of her career working in Alaska. The jobs described in this book take on a special challenge when they are performed in the wilds of Alaska, "the last frontier." Travel is often by plane or the boat you navigate yourself. Working in grizzly bear and moose country adds an element of potential danger not encountered in many places. Profoundly beautiful surroundings richly reward those who choose this rather solitary life in remote locations. Subsequent chapters include the stories of other women who spent part or all of their Forest Service careers in Alaska.

Usually a core team of specialists works together on interdisciplinary teams (IDTs or ID Teams) to analyze potential effects of proposed projects (timber sales, mining operations, stream restoration, recreation area development — any action proposed on national forest lands) on the environment. Individual specialists also plan and implement projects to improve their specific natural resource area, such as wildlife or fisheries habitat improvement.

It is not uncommon for employees to have held positions in more than one specialty area while progressing in their careers. The Forest Service offers a lot of training, including "cross-training," which allows employees to gain experience in a variety of areas. Besides enhancing career choices, a diverse understanding of related disciplines helps those working on ID teams to understand each others' perspective, strengthening the cohesion and hence the productivity of the team.

The ranger district level is considered to be closest to the ground, compared to the forest, regional, or national levels, which are increasingly administrative and political. Most work for district specialists occurs in the field through the GS-9 level. Beyond that, a fair amount of time is devoted to indoor administrative duties. Professionals with administrative duties at the district level may still spend as much as fifty percent of their time in the outdoors.

The biographies in this chapter offer a sampling of work performed by professional employees at the ranger district level.

∼

Jane Bard, Silviculturist

Jane Bard (Figure 26) grew up in a suburb of Pittsburgh, Pennsylvania, on a cul-de-sac that provided a lot of street and yard area, as well as a few little woodlots that afforded a lot of imaginative outdoors play, including playing in a tree house.

Jane's family did not do outdoor activities together; her parents both worked and visits to family members and picnics were their only recreational family activities. Her parents had hobbies like picking blackberries, canning, crocheting and building things.

As Jane and her friends got older they played in the woods and rode bikes. Jane also read, played the piano, cooked and sewed. In school, art was her favorite subject. A "great female chemistry teacher with a prosthetic leg"

Figure 26. Jane Bard.

helped Jane feel competent in science and shared her admiration for an important female scientist, Marie Curie.

Like many of us, Jane feels that she fell into her forestry career by accident, but travel and being in the woods attracted her to it, then and now.

She has a Bachelor of Science degree in forest management from Penn State. As a student, she worked as a seasonal Forest Service employee. She took a few non-degree graduate classes while waiting for her husband to graduate, and worked part time at a research lab. She and her husband went into the Peace Corps in Honduras, and worked in a watershed research and restoration project in a small town near the capitol. Jane considers the two years they spent in the Peace Corps to be her most extraordinary experience. They recorded erosion, runoff and weather data, maintained and built erosion control structures and built some terraces and weather stations. They worked with a Honduran forestry technician as their counterpart, and led a crew of six other workers. Jane said, "It was a type of travel completely different from going on a vacation, and it was fun. When it wasn't frustrating, sad or hard. It was full of vivid memories, vivid people and vivid experience. Just like the ad used to say — the toughest job you'll ever love." Non-competitive eligibility, an advantage given to those who serve in the Peace Corps helped Jane get her first permanent Forest Service job.

After returning to the United States, Jane got a job as a special uses, lands and minerals forester on the Dixie National Forest in Utah where she and her husband lived for three years. Since then, she has worked as a timber forester on the Monongahela National Forest in Richwood, West Virginia, closer to their families. Jane would have liked to move more for advancement or just to see more of the world, but decided against it because her husband preferred not to move. It has been a challenge to maintain her creativity and productivity, not get in a rut and always find some new interest in the same town and the same job.

She has worked as a silviculturist for many years, but has also had a few details into different positions, such as NEPA

Figure 27. Jane examining first year regeneration success in a recent clearcut.

coordinator and acting district ranger. She now has over thirty years of federal service, and is eligible to retire, but she enjoys this "land management hobby" too much to consider retiring right now.

As a silviculturist, Jane's job has periods of being mostly in the field (Figure 27) or mostly in the office.

She writes environmental documents or prescriptions or analyzes data, or sometimes supervision and planning take up a lot of time. At other times she evaluates potential timber sales, sprays gypsy moths, girdles trees or supervises contract workers (Figure 28).

At this time in her career she works a lot on her own, so she is not tied to others' schedules. Long days have often been part of her daily schedule. She has a passion for doing what she can to maintain and enhance the diversity of the Appalachian forest. She works on that by minimizing deer damage to young trees, planting rare species like butternut and hybrid

Figure 28. Jane checking for hemlock and beech mortality.

Figure 29. Jane planting hybrid American chestnut trees, Williams River area.

chestnut (Figure 29), being watchful for sensitive plant species or non-native plants and educating others about forestry. She noted that in the last few years, as the experts she relied on for advice and answers have retired, realizing that she is now the "expert" is extraordinary and unnerving. She said, "I need to keep stepping up, every day."

Jane did not deliberately choose a non-traditional career, but thinks she must surely have noticed how few girls were in her classes in college. She found the subject matter interesting, and says that before she knew it she graduated with a degree in forest management, and some related summer and part time work experience.

She is married, *with children*. Her husband worked in forestry for a while, but for quite some time he has been a stay-at-home dad. She said her second most extraordinary experience was, "the day I stopped moving long enough to realize I was really supporting a family of nine, and had been for several years . . . that was extraordinary and a little unnerving." As of this writing, two of their seven children were still at home, one in high school and one in middle school. Their oldest is over thirty, the youngest had just entered her teens. The five who have left home still take up a lot of energy and time, just to keep in touch. Some are still in West Virginia, and some are fairly close in Pennsylvania.

A final noteworthy experience Jane shared was the day she passed the pack test — a rigorous physical test required for fire fighting — for the first time. She stopped at the forest supervisor's office on the way back from the test location, and felt as though everyone there already knew she had passed and they were happy for her. She proudly remarked, "As an over-forty, fairly new mother of seven who had trained very hard, passing the pack test that time was a truly extraordinary experience."

∽

Kathy Burnett, Wildlife Biologist

When she was young, Kathy (Figure 30) played in the fields among the Brussels sprouts surrounding her family's home in Salinas or the cotton fields surrounding her grandparent's home in the central valley of California. She rode her bike through the fig orchards that surrounded a new tract home in Fresno, California. Sadly, the fig orchards are all gone now. Kathy has memories wandering their gravel road with her wagon full of water, collecting earthworms after rainstorms in Salinas. As a teen-ager her parents took the family to Bass Lake in the Sierra Nevada Mountains north of Fresno to camp for two weeks, and then left the kids with friends for another two weeks, so they could spend weeks every summer playing in the woods and water.

Kathy does not remember having an affinity for math and science until high school, when she enrolled in an advanced biology course and was influenced by an instructor who taught hands-on how to conduct scientific experiments, engaging students in real-life situations. After that she gravitated to the high

Figure 30. Kathy Burnett.

school biology classes, and then continued in college, taking every natural history, biology, botany and ecology class she could fit into her schedule, first at Columbia Junior College then at UC Santa Cruz, both in California. She developed relationships with a number of mentors throughout college and her career. She was introduced to field classes at Columbia Junior College, taking one memorable field class to the White Mountains with Ross Carkeet, a forestry teacher, and others with Ken Norris from UC Santa Cruz. She took plant identification and plant geography courses from Ray Collett at UC Santa Cruz. He had them make line drawings of wildflowers, to learn botanical identification terms. Another UC Santa Cruz instructor, Barry Hecht, taught her field data collection methods.

When Kathy moved to the Sierra Nevada in 1976, after she graduated from college with her BA in biology, she wanted to continue being in the field and thought she would like to work for the Forest Service. She visited the Mi-Wok Ranger District on the Stanislaus National Forest and talked to the district ranger, asking about jobs in biology. He told her that the Forest Service hired mostly foresters not biologists, and suggested she apply for clerical jobs. She did, and ended up as an admissions clerk at Columbia College. She spent her free time bird-watching every week-end, walking through the forest and sitting by a pond, learning the Sierra Nevada birds by sight and call.

The following year Kathy landed a two-month temporary job supervising teen-agers at a live-in Youth Conservation Corps program. Then she was hired into a permanent clerk-typist position on the Groveland Ranger District on the Stanislaus National Forest. She had her foot in the door, but in 1978, getting from a clerical job to a job in biology was a challenge. The agency was just starting to hire wildlife biologists, and fewer botanists. Going from a clerical series to a professional series required a conversion, which Kathy found out was a rare thing. Luckily the district wildlife biologist, Dennis Danner, was willing to hire her as a temporary biological technician for the summer. She jumped at the chance, giving up a permanent position to finally work in the field. The district ranger counseled her to stay in her permanent clerical position.

Kathy said, "That was 1979, and to give him credit there really weren't many women in field positions, so we were unproven. But did we ever have to prove ourselves. You didn't dare call in to get help if you got stuck, you figured a way to get un-stuck. You kept up with the man who would lead you straight up a hill without stopping. At the time I weighed 105 pounds and I worked as hard as the men and didn't complain." The sense of having to prove herself persisted for twenty years into her career before she felt she truly belonged. She said, "In the 90s, I finally felt I'd proven myself enough and started to push back on any man who was silly enough to say something."

Through 1989, Kathy worked a number of Forest Service temporary jobs, progressing from GS-5 through GS-9, as a gardener, biological technician, botanist and wildlife biologist; this, after going back to graduate school at UC Davis to take classes to qualify for biological science. In between Forest Service temporary jobs, she worked as a consulting biologist, keeping her botany and biology skills up-to-date by doing botanical and wildlife field surveys in the private sector. Jim Frazier, Forest Hydrologist, mentored Kathy throughout a large portion of her career. She says "he gave me sage advice over the twenty-five years I worked with him on projects."

Finally, in 1989, the stars aligned. Kathy was working for the forest biologist, Tom Beck, as a temporary GS-9 wildlife biologist supervising volunteers doing restoration work after the Stanislaus Complex fire. She also started doing some wildlife input work for the Summit Ranger District projects, because their district biologist position was vacant. She remembers Tom Beck taking her to the Summit Ranger District and introducing her to Herb Hahn, the district ranger. They sat and talked about the projects Kathy was working on and other things. She says that looking back, it was the most informal interview she had ever experienced. Thanks to Tom Beck's recommendation, the support of Herb Hahn and the push to hire women due to the consent decree, she got her second permanent job, this time a professional appointment to a GS-9 journey-level district wildlife biologist position. She declared, "Finally two feet in the door!"

Kathy worked as the district wildlife biologist during the spotted owl controversy for eight years, nearly burning out. She was actually considering resigning, this after taking so long to get her two feet in the door! But with advice from regional botanist Jim Shevock, and help from her district ranger, Karen Caldwell, she made another transition in 1997. Jim helped by giving the advice to stay in the Forest Service, but move off the "front line." Karen helped by supporting Kathy's idea to get a Master's degree, and being flexible with her work schedule. It took three years, a Master's in human resources and a forest re-organization, but Kathy moved into the administrative job that she held until her retirement. She supervised visitor information, interpretive services and administrative staff and worked with partnerships, agreements, hydropower coordination and public relations and communication.

A major highlight of Kathy's career includes riding in the cab of a D-8 Caterpillar tractor with the operator, driving through a sea of 20-foot tall old-growth chaparral, crushing it flat and putting in fire lines so they could do the first prescribed burns on the district. The prescribed burns were ten to twenty acres in size and were experimental.

Conducting ten to twelve hour overnight spotted owl surveys in the wild country of the Groveland Ranger District, before there were established protocols was another memorable experience. At the end of one shift she visited the Trumbull Peak Lookout at dawn and watched the sun rise over Yosemite Valley, Half Dome shining in the sun.

Yet another extraordinary experience was hiking over "Big Sammy" from Leavitt Lake to visit historic check dams in the Emigrant Wilderness. The view from the top of the pass, looked down on High Emigrant, Middle Emigrant and Emigrant Meadow Lakes. It was an hours-long descent into those lakes from the crest of the Sierra Nevada.

Kathy is retired now and still holds a passion for hiking the mountains. She takes frequent long hikes with groups she has joined, and accompanied by her dogs (Figure 31).

She hopes for younger generations to discover the joy of being in nature. The Forest Service has a Kid's in the Woods program to get kids back into the outdoors. Kathy feels lucky to have had a connection to nature when she was growing up, which she thinks pulled her into her work with the Forest Service. She mused, "We sometimes joke that in the future we won't be doing field work, we'll be at a computer piloting a drone, or interpreting satellite photos, or using Google Earth to plan our resource projects. Let's hope that doesn't happen. The experiences and adventures you get from working for the Forest Service are hard to replace with a computer screen and a joystick."

Figure 31. Kathy hiking the Sierras with her dogs, left to right — Cleo, Ray Charles and Salty.

∼

Anita Leach, Forester

Anita Leach (Figure 32) was born in San Luis Obispo, California and grew up in Mountain View and Sunnyvale, California. She currently lives in Mill City, Oregon.

Anita loved to camp and hike as a kid. She also loved any kind of sport, especially tennis, volleyball, softball and swimming. She liked being outside and playing with her friends. She also loved to read at night when she had to go to bed. She said, "If we read we got to stay up for an extra hour. If we

Figure 32. Anita Leach.

didn't read, it was lights out right away." She disliked having to stay in the house after dinner and be in bed in the summer before the sun went down. She also disliked liver and lima beans. Her favorite school subjects were science, math, sewing and physical education.

Anita's family values included close family relationships, strong friendships, responsibility and accountability, honesty, hard work, achievement, helping others and having integrity. Her family enjoyed weekend hikes and picnics as well as annual vacations, which almost always included camping, hiking, fishing, boating, water skiing and swimming. They had to spend one day each weekend with just their family.

Anita was attracted to forestry because she loved the outdoors and the woods. She loves the variety in her job and the fact that, "I can go out in the woods and actually get paid to do it." She loves working with people and trying to solve problems together. She said, "There are lots of challenges but that is also what makes it interesting."

Sometimes she goes to the woods, and sometimes leads IDT meetings, makes GIS maps or writes environmental documents. Other times she leads public meetings.

Anita got her Associate Degree from West Valley Junior College in 1975, and her Bachelor of Science in forestry from Humboldt State University in 1979. She worked seasonally during college from 1977-1979 on the Snow Mountain Ranger District of the Ochoco National Forest in Burns, Oregon, doing stand exams, superior tree selection, pre-commercial thinning contract inspection and fire fighting.

After graduation she was a reforestation forester for the Detroit Ranger District of the Willamette National Forest in Detroit, Oregon. She did planting survival surveys, planting and pre-commercial thinning contract inspection, stand exams, selected superior trees, collected seed, prepared contracts, kept records for field surveys and completed annual accomplishment reports.

Anita stayed on the Detroit Ranger District until 2004. That time was marked by promotions and details that expanded her skill set. The Forest Service offers opportunities to cross-train, and to gain multiple skills as an employee follows their career path.

As a supervisory forester from 1979-1980, Anita supervised a seasonal crew and worked in pre-sale. She inspected a salvage sale for contract compliance. She switched to fuels management from 1981-1982, doing project planning and work planning and coordination for 110 seasonal employees doing various brush disposal projects.

Anita then worked as a planner for several years, with a two-year stint as the acting district wildlife manager sandwiched in. As a planner she led interdisciplinary teams, took projects through the NEPA process and wrote environmental documents.

She took another detail as acting integrated resource planner overseeing the budget, and then a special assignment as a forester, analyzing the district land base and resources to determine timber sale volumes the district could support while meeting Forest Plan standards and guidelines.

In 2004, Anita transferred to the Sweet Home Ranger District on the Willamette National Forest in Sweet Home,

Oregon, as their planner, which remained her job until she retired in 2017. She commuted from her home in Mill City. She continued to broaden her experience with details, including natural resource planning staff officer on her forest, environmental coordinator on the Prairie City Ranger District of the Malheur National Forest and forest environmental coordinator on the Willamette National Forest.

Between 1999 and 2007, Anita also served as the project director of the 21st Century Community Learning Center, Santiam Canyon School District, in Mill City, Oregon. There she led the academic and recreational after school and summer program for students in grades K–12. The program offers adult parenting, and computer and art classes. Anita marketed programs, wrote grants, supervised employees, managed budget and taught earth science, natural resources and art.

Anita did not deliberately turn to the non-traditional, but growing up she had two brothers and almost all of her cousins were boys. If she wanted to play with them she had to be able to do what they did, and many of them were older than her. She worked really hard to keep up with them. She also greatly admired her grandfather. She speculates that this mix may have led her to this career.

There weren't many women in the woods when Anita started with the organization and there was one man who told others that he was going to make her quit. His idea was to take her out to a really steep unit and work her so hard that she would quit. They got to the unit and Anita got out of the truck to put on her rain gear and was expecting a briefing. But the man just got out of the truck and took off without telling her where he was going. She looked over and heard him crashing through the woods. At that point she said, "I was bound and determined to keep up with him if it killed me. I had to listen for his movements and watch the brush move, but I followed him and caught up. I got stung and scratched up by the brush but I wasn't going to let him know that."

Anita cited the most extraordinary experience of her life as the birth of her children. She was a working mother, torn between her kids and her job. She compromised by working part-time when they were young and spending quality time with them when she was home. She was diagnosed with cancer at age thirty-nine and went through surgeries and treatments while raising her two young children and working. The children are now grown, married and on their own. She has truly enjoyed being a mom and nurturing her kids, along with her husband of over thirty-four years.

What drives Anita now is to try and do her best at whatever she does and to continuously improve. Her passion is giving back to her community, reflected in her greatest goal to be a good and caring human being.

∼

Lynda Perry Mills, Wildlife Biologist

Lynda (Figure 33) grew up in a Forest Service family. When she was born in Arizona, her father was working at a job corps center on the Apache-Sitgreaves National Forest. When she was five, he was transferred to another job corps center on the Cherokee National Forest and when she was eleven, he took a job at a job corps center on the Daniel Boone National Forest in Kentucky. At this same time, his father, Lynda's grandfather, was finishing up what would be a 36-year career with the Forest Service as a fire control officer on the Daniel Boone National Forest. In spite of the fact that Lynda became the third generation in her family to work for the Forest Service, she says that having a career in the Forest Service never crossed her mind as a child. When she did contemplate the kind of career she might have one day, two things were foremost: she did not want be in a classroom all day as her father had been, and she did want an outdoorsy job like the one her grandfather talked about. However, because they both had such different jobs, it was not until she became an adult that she made the connection that both her father and grandfather were working for the same agency.

Lynda considers Kentucky her home. Coming of age in a sparsely populated, largely undeveloped and highly scenic Appalachian community in southeastern Kentucky had a

Figure 33. Lynda Perry Mills.

significant influence on her. With little else available for entertainment, her family put great value on being outdoors. During the week, they spent a lot of time outside working in their garden, cutting firewood, bush-hogging, mowing and clearing brush. For enjoyment, they would often go hiking or fishing, usually on Sunday afternoons. Both of her parents enjoyed being outside and living a country life.

It was natural that as a child, Lynda loved being outdoors and played outdoor games like kickball and hide-and-seek with the neighborhood kids. She hated housework and cleaning, but liked cooking. She enjoyed helping in the garden and doing yard work. She helped tend a flock of chickens for five years, and the pet dog was her responsibility. She and her dad liked to squirrel hunt together, and the two of them frequently helped her grandfather with outside work while her sister and mother helped her grandmother with inside work. She spent all her free time outside and could never get enough of it. If she was not reading a book, doing homework or her indoor chores, she was outside. She jokingly tells people she was "my father's son." All the typical "boy things" a father expects of a son, Lynda did. She stacked wood, dragged brush, mowed grass, worked in the garden, helped with the carpentry jobs, fed the chickens and learned to hunt and fish. She said, "I always thought boys got to have all the fun. They were always moving and playing outside. The things they were encouraged to do were the things I also wanted to do. As a child, I had a hard time sitting still, especially indoors, but boys were expected to be restless and were encouraged to go outside."

Lynda loved reading, English, writing and history in school. She hated math classes and chemistry but liked physics and biology. She took a shop class in middle school and was the only girl in class. She wanted to be in JROTC in high school but was discouraged from that by the teachers and by her school counselor. She belonged to 4-H for one year and Girl Scouts for twelve years. In high school and college, she especially liked the hands-on aspects of biology, such as dissecting frogs and going on outdoor field trips to learn about plants and ecology.

Although she did well in school, she hated every moment of it. She felt bored much of the time and sitting in a classroom all day felt like prison. She counted the days until she could graduate and start making her own decisions. When the time for college came, the thought of spending four more years sitting in a classroom felt torturous. Still, she was excited about the idea of getting to choose her field of study. Her father had strongly encouraged her to study something either in math or science because "that's where the jobs are," but she knew math and chemistry were not her strong suits. So instead, she focused on the one science class she did enjoy — biology. As she thumbed through the biological sciences section of her college catalog one day, she ran across the "Wildlife Management" curriculum. The courses included subjects like ornithology and mammalogy. Field studies in wildlife and botany jumped off the page at her. "Wow," she thought, "those classes sound interesting, plus I might get to go outside! Sign me up!"

Lynda announced her grand plans to her parents. "I'm going to study wildlife management!" she exclaimed, proud that she had made a decision on her own and sure they would be pleased to see that she had chosen something in science. However, they were less than excited. "We think it would be better if you got a biology degree with a teaching certificate," her father suggested. If there was one job she knew she did not want, teaching was it. She wanted a profession that would get her outdoors at least. She says now, "That is the kind of career I'd found and that is what I still love about it!"

In hindsight, she said that her family's' reaction to her planned course of study was her first inkling that she had chosen a "non-traditional" career path and that it was not going to be easy. "And it wasn't," she remembered. As a woman in college studying wildlife management, she was always the minority in the wildlife classes and several of the science classes. One of the first challenges she faced was having professors in wildlife science who were less than encouraging. She recalled one incident where she was directly confronted by a wildlife professor. She was working on a scholarship application and asked him if he would be a reference for her. He had served as her advisor for two years, and she had been in his classes, with a good grade average. So, she was completely surprised when he refused to let her use him as a reference. When she asked why, he told her point blank that he was "not going to help me anymore because he didn't believe women should be in wildlife management." Lynda was stunned. She had not pegged him for being sexist. As her advisor, he had been less than helpful and encouraging when it came to planning her courses and schedule, and in his classes, she often felt that he nit-picked some of her answers on exams, but she had never put two and two together until then. What she had thought was him trying to "help me help myself" was really just his belligerence because she was a woman! She expressed her disappointment and found another professor who was genuinely supportive for her reference. She says now that he taught her a valuable lesson in that she learned just how well others can hide their true opinions regarding women in non-traditional careers.

Lynda graduated *cum laude* from Eastern Kentucky University with her degree in wildlife management in May 1991. She started her career with the Forest Service the following month, with a temporary one-year appointment as a GS-5 wildlife biologist on the Daniel Boone National Forest. She was stationed on a ranger district located a short distance from her hometown. Before that one-year appointment had expired, Lynda was placed in a permanent wildlife biologist trainee position on another district located a little bit further from her hometown. At that time, trainee assignments required the employee to sign a mobility agreement so that they may be placed in an available permanent position anywhere in the agency at the end of their two-year training period. Lynda trained for her two years under Steve Phillips, the district wildlife biologist. Steve took Lynda's training seriously and she considers those two years one of her best learning experiences and credits him

Figure 34. Lynda threading mist net for bat survey.

Figure 35. Lynda and coworker setting up bat mist net.

with giving her the skills she needed to succeed as a district biologist on her own. As fortune would have it, just as Lynda's training period came to an end, another biologist on the Daniel Boone decided to resign from the agency. Lynda was transferred into that vacant position, which just happened to be the district on which she'd grown up. After three years, she had already come full circle and was back home again. While there, she was promoted to the journey-level GS-11 grade, and she worked as that district's biologist until transferring to the Mark Twain National Forest in Missouri in June 2002.

Whether on the Daniel Boone or Mark Twain National Forest, Lynda's days varied a lot. She spent a lot of time writing reports and reviewing projects for potential impacts upon wildlife species, particularly endangered and rare species. This required her to know the habitat and life history requirements of the species, so she read and learned about those plants and animals that she had to consider. She also conducted wildlife surveys to determine presence/absence of some of these species, such as endangered bats (figures 34, 35 and 36). She worked with botanists to conduct plant surveys of project areas. She conducted breeding bird, mussel and amphibian surveys.

Figure 36. Lynda and colleague setting up mist net for bat survey.

During the first half of her career, she primarily worked with red-cockaded woodpeckers and developed habitat improvement projects for them and monitored their populations. She learned to handle and band the birds, and coordinated and assisted with inter-state translocations. One of her most extraordinary career experiences came on December 5, 1994. At that time, red-cockaded woodpeckers had not been seen on her district in over ten years, and Lynda had been working for the previous two years to improve red-cockaded woodpecker habitat there in hopes of attracting them. She recalled with fondness that chilly day in December when she walked into a shortleaf pine stand on her district and looked up to see a red-cockaded woodpecker calling from the pines above her. Later that evening, she returned to the same site and watched as the woodpecker roosted in one of the artificial cavities that she and her technician had developed. That moment changed her focus and convinced her that "if you build it, they will come!" For the next seven years, Lynda focused on expanding the red-cockaded woodpecker population on her district and the population continued to grow until its premature demise in the summer of 2001, due to a southern pine beetle epidemic. After spending a decade focused on trying to save the species, witnessing the rapid and unforeseen loss of the entire red-cockaded woodpecker population in the state of Kentucky was a devastating moment in her career. Before reporting to the Mark Twain National Forest, she took several days of leave-without-pay so that she could write a paper about the red-cockaded woodpecker extirpation in Kentucky, which was published in the 2003 *Proceedings of the Red Cockaded Woodpecker Symposium*. She said, "I knew that in another decade or two, there would be no evidence left that indicated red-cockaded woodpeckers had ever existed in Kentucky, so I thought it important that someone document the events that preceded their extirpation before they became a distant memory."

Regardless of what project she was working on at the time, her work always required her to collaborate with a wide variety of others such as hydrologists, soil scientists, archeologists, foresters and fisheries biologists, to develop and plan projects that benefited wildlife and other natural resources. She enjoyed learning from the other disciplines in the agency and working together to develop projects that benefited entire ecosystems rather than just a single species. When she worked on the Mark Twain National Forest, she became more involved in recovery efforts for endangered bats. Noting the lack of good survey information on the national forest, she and a few other biologists decided to become trained in handling and identifying bats, and their efforts contributed to the overall knowledge base for species such as Indiana bats, gray bats and more recently, the northern long-eared bat. In the last few years of her time on the Mark Twain, Lynda also took an interest in wetland restoration work. Training under Tom Biebighauser, Lynda learned about wetland delineations and began to develop projects on the national forest to restore and rebuild small wetland habitats. She particularly enjoyed working with hydrologists, soil scientists and botanists to develop projects that benefited not only wildlife species but restored hydrological systems as well.

After working for a few years in her career as a wildlife biologist, Lynda felt very satisfied with the career choice that she had made, but she does admit that when she deliberately chose to enter wildlife management, she underestimated the number of personal challenges she would face by being in a non-traditional career. She would have preferred that more women were in her field when she entered it, and she is glad to see more women and other minorities entering these field positions now. Not only was she a minority as far as gender, but she was also hired during the time when professionals such as wildlife biologists versus foresters were new to the agency. As a biologist, her philosophy came across as being very different than that of the majority. Many of her fellow co-workers assumed that because she was an "ologist," she was anti-timber and anti-management, but she was not. She learned early that she had to earn the trust and respect of her peers on many different levels.

It was a big challenge gaining the respect of her male coworkers. She said that women often were not taken as seriously, and frequently did not receive as much backing from their supervisors as the men did. Many men resented women for taking jobs that they thought the men deserved and were more qualified for, and in some cases, that was probably true.

Lynda said that the biggest challenge was learning to work independently with little camaraderie in the workplace. She said, "Unless you primarily worked in the office, women in the agency tended to not have 'buddies' to work with and support them as much as the men did. The women tended to have to work alone and didn't have social circles in the workplace as much as the men did. As a matter of fact, the only question that was asked of me before I took my first job with the Forest Service was 'did I mind working in the woods alone?' Oddly enough, it was a woman who asked me that."

With time, Lynda realized that subtle sexism also came with the territory. In meetings, she sometimes found that when she tried to add to the discussion, she would be met with silence or "talked over," but if one of the guys in the meeting said the exact same thing, he would be encouraged to elaborate. One male supervisor would make a point to yawn when Lynda started talking during meetings, or would wait until she started talking and use that as his chance to go to the restroom. She said it sent a message to all of his employees that it was okay to not take her very seriously.

She also wanted to do more in the fire organization but lost her enthusiasm for that after a situation occurred while she was a squad boss. She had been leading a squad on a wildfire when one of the rookie male members of her squad refused to follow her instructions to stop firing the line. His refusal to stop firing put her entire squad at risk. Later, when she brought this matter to his attention, he became belligerent and obstinate and laughed at her. When she brought the matter to his supervisor's attention, nothing was done about it other than she was told that they would no longer put this guy on her squad. This influenced her decision not to pursue higher fire positions because she did not feel that her directions on the fire line would always be followed and she was concerned that it could put her and others' lives at risk. She did not feel like she would have any backing should something go wrong.

In spite of a few negative experiences, Lynda described her overall experience working with men in the agency positively:

> Of course, there are many men who do support women, and you mostly ignore or avoid the ones who don't. And you don't hold a grudge because it will just make you bitter. Some things just won't change overnight but in the 20 years that I was there, I did see some substantial changes for the better.

Lynda knew other women who struggled with discrimination but most did not leave the agency. Many simply chose to stay in the office more rather than work alone in the woods, or they chose another career field within the agency that was more "female-friendly." She said she stuck it out because she is stubborn plus she felt confident in her abilities. She also had a lot of support from her father, a handful of great co-workers and a very good (male) mentor within the agency. She said, "It was the men who supported and encouraged me who really inspired me not to give up. Their support made me feel like there was hope."

She added, "I am such a better person for my time in the Forest Service. It made me stronger physically, emotionally and gave me much more confidence. I learned valuable skills and worked with a level of independence that I wouldn't have likely found elsewhere. If you can hang in there, you'll eventually find yourself stronger and more confident than you ever thought you could be."

In August 2000, Lynda and her sister tagged along with Lynda's brother-in-law on a business trip to Melbourne, Australia. While there, Lynda met Tom Mills. Tom was living in Melbourne at the time and was a friend and colleague of her brother-in-law. During their visit, Tom served as their host and tour guide. Lynda said, "I fell for Tom when I noticed he had a bird book on his coffee table. Turns out, he was really into birding and he taught me lots of Aussie birds during that trip. The rest is history...."

Tom and Lynda maintained a long-distance relationship after meeting, and they got engaged in April 2001. They tossed around where they would live after they married, and settled on Tom coming back to the United States.

Tom resigned from his position in Australia in September 2001, moving to Kentucky where Lynda was working, and they were married three weeks later. Lynda laughingly said, "We had to get married — the long distance relationship was costing us tons!" They lived in Kentucky until Tom accepted a job with a small consulting firm in St. Louis, Missouri in June 2002. When Lynda told her supervisors on the Daniel Boone National Forest that she would be resigning so that she could move with her husband, she learned that there was a Forest Service ranger station near St. Louis that needed a biologist, and she got a job there. In July 2002, she went to work full time as a wildlife biologist on the Potosi Ranger District of the Mark Twain National Forest. She and Tom lived near St. Louis, each commuting about thirty miles in opposite directions.

Nine years into their marriage, Tom and Lynda became parents. Lynda gave birth to their son in July 2010, and soon after, they adopted another son when he was born in February 2011. At that time, Lynda was close to having twenty years of service in the Forest Service, which would vest her for retirement. She continued working until she completed that twenty years, then she chose to resign to be a stay-at-home mother. At the same time, her parents were getting older and needed a family member close by. That dovetailed nicely with the fact the cost of living was lower in Kentucky than in St. Louis, so Lynda and Tom decided to move back to Kentucky, to the same ranger district where she had worked a decade before. Tom's supervisor allowed him to telecommute, so, Lynda said with a smile, "We moved back to my home town and tightened our belts. We now make ends meet on one income, homeschool our children and grow potatoes." Once again, her life had gone full circle and she was back where it had all begun.

Figure 37. Lynda and family at Cumberland Falls, Kentucky.

Lynda worked for twenty years as a wildlife biologist for the Forest Service and she said she loved nearly all of it. For years, she could not think of herself doing much else. She said of those years, "I never felt the pull to become an administrator, nor to follow another career path. I occasionally thought of leaving the agency, but never the wildlife profession. I feel as confident today as I did thirty years ago when I made that fateful decision to study wildlife management."

Lynda pondered what her true passion is and commented that she can become "too passionate" regardless of what she is doing. She laughingly said it goes with her overachieving personality. Now that she has chosen homemaking as her current vocation, she and Tom have adopted again and are considering fostering other children.

She admits to being a perfectionist at times and believes that "if something is worth doing, it's worth doing well." She says she has never worked just for a paycheck but instead, tries to measure success by the level of commitment she puts into something, and whether or not she helped make things a little better than they were. She said, "Too many people consider their work 'just a job,' and get mired down in hopelessness and apathy." She hopes she has helped inspire people she has worked with over the years to look beyond the paycheck and see the joy that comes in simply trying to do a job well.

Lynda's love of wildlife and the outdoors has taken her on some great adventures, to some beautiful places and even led her to her spouse. It is a part of her that will never die. Today, she still volunteers during the summer on national forests with bat surveys, and does occasional outdoor education programs for the public. She and Tom still go birding together. But most of all, she enjoys sharing her love of nature with her children (figures 37 and 38). Like her, they are growing up in the national forest.

Lynda keeps the option open to return to work when her children are older, with the Forest Service, another natural resource management agency, or private consulting. She said, "I haven't ruled anything out. I am waiting to see what happens next."

Figure 38. Lynda and boys at Cumberland Falls, Kentucky.

When asked if she thinks one of her children might follow in her footsteps and become the fourth generation of her family to work for the Forest Service, she laughed and mused, "I bet it will never cross their mind."

∼

Wendy Parker, Archaeologist

Wendy (Figure 39) was born in Carmichael, California and lived in Orangevale, California until she was eight when her family moved to Oroville, California, where she has lived since, and has managed to have a career with the Forest Service, in place.

Wendy grew up playing in the outdoors. Her family took annual trips across the United States, Mexico and Canada. She spent her summers traveling, camping, fishing and enjoying the outdoors. Her family trailer-camped all across northern California and the Desert Valley. They were active in the California Snowmobile Association and spent most of their winter weekends snowmobiling in northern California. Her dad

Figure 39. Wendy Parker.

took her fishing at every opportunity, up until she started high school and chose not to go.

In school her favorite subjects were history and social studies. She disliked organized sports and physical education; She was more into riding her bike and hiking. She described herself as a curious person who loves learning about people in general.

The Public Broadcasting Channel aired an archaeological program geared toward children, which captured her attention at a young age. Throughout elementary school she had many teachers who exposed her to different aspects of archaeology and she always excelled during those portions of the class. She still remembers to this day every archaeology curriculum she took part in during her elementary school education.

She took every anthropology class offered at the local Butte Community College. Even with her earlier exposure she had never realized that she could actually gain employment within California in archaeology. She eventually transferred to CSU, Chico as an anthropology student and was exposed to many California job opportunities. She worked for the Archaeological Resource Center at the college who connected her with her first seasonal position with the Forest Service.

Wendy said she loves being able to learn about different ways of life and the history of the area in which she lives. She said, "I love being able to share the knowledge that I have with both the public and other Forest Service employees. A large part of an archaeologist's job is to share knowledge and we take every opportunity to involve the public in our work and it is one of the most rewarding parts of my job. Watching young people learn and get excited about archaeology is one of the best parts of my job."

Wendy started as a seasonal archaeological technician, moved into a Student Career Employment Program (SCEP) position and converted to an assistant district archaeologist. She managed seasonal crews in accomplishing archaeological surveys and site recording for project work and the data entry associated with the work. She managed the district's archaeological artifact and historic document collections. She conducted National Register of Historic Places evaluations and served on an IDT providing input on how to protect archaeological resources during project planning and implementation.

She has developed and polished her database management skills, and is a Natural Resource Management (NRM) employee who assists other archaeologists with database entry problems. NRM is a system of database tools for managing agency data across the Forest Service. It includes the INFRA database used to manage infrastructure data. In 2013, Wendy joined the National Heritage Information Management Initiative Team when she took on heritage helpdesk roles for the INFRA reporting database within the natural resources management program. She works part time assisting the nation's Forest Service archaeologists with program and data entry questions and concerns.

She took on the role of Region 5 Heritage Data Steward in 2014. She serves as the contact for the regions' archaeologists on all matters of archaeological data and end of year reports. She ensures that the regions' archaeological data is entered correctly into the national INFRA reporting database within the natural resources management program. She trains the regions' archaeologists on INFRA. She assists in quality control in generating yearly reports that the regional archaeologist sends to the California State Historic Preservation Officer. In 2014, she joined the National Resources Management Modernization Team to assist in the development of a new heritage database. She helps ensure that heritage business rules are followed in the development of the new system and helps ensure that Region 5 data migrates into the new system as it is developed.

In 2015, Wendy accepted a detail to the adjacent Tahoe National Forest. The Tahoe received a large settlement from a fire and a team was brought in to use the funding to restore the landscape. Wendy leads that team. She oversees the NEPA preparation and implementation of ecological restoration work, such as fuels reduction, mastication, under burning and hardwood culturing within and adjacent to the fire footprint. She developed the program, from securing a building for the team to work in, securing fleet, to leading the IDT, to running the contracts as the contracting officer's representative. As part of this she also runs an archaeological crew conducting surveys, resource protection and National Register of Historic Places evaluations.

Wendy had not deliberately sought out a non-traditional job, but followed her heart and found a job in her hometown doing, as she said, "What I loved, and I jumped on the opportunity to work close to home. Archaeology is traditionally a job that requires a lot of traveling and living away from home and I wanted to stay close to my family while building my career." Wendy and her husband live less than five miles from the Forest Service office where they both work. Both of their immediate families live less than two miles from them. She was able to accept her assignment on the Tahoe National Forest without having to relocate. She now commutes about forty-five minutes each way to her worksite, or telecommutes from the Plumas.

Wendy balances her desire to have a career based in her hometown with a passion for traveling. She says she is happiest when she is traveling and experiencing the different cultures of the world. True to her chosen profession, most of her travels have taken her to many archaeological sites around the world. She has been all over the United States, Mexico, Belize, Guatemala, Honduras, Peru and the Galapagos Islands. She loves seeing, learning about and photo-documenting her travels. One of the most memorable of those experiences was climbing to the top of Huayna Picchu with her husband. She said the experience was surreal, climbing to the top of the 8,924-foot elevation mountain and looking down on Machu Picchu and its magnificent surroundings.

She is the first of her family to obtain a degree, and subsequently an advanced degree. She had to support herself and pay for her entire education; however she feels she was fortunate to be able to take advantage of opportunities to be employed as an archaeologist while attending school.

Wendy's job supports her greatest passion of being able to create opportunities for the public to be involved in archaeology on the forest she works on. She loves creating public interpretation, and designing information to be dispersed for public use and interaction. She appreciates that performing her job often allows her creative juices to flow, using her graphic design and photography skills.

∼

Cindy Roberts, Wildlife and Aquatic Biologist

Cindy (Figure 40) was born and raised in Grafton, North Dakota, a small town of less than 10,000 people, considered one of North Dakota's moderate sized "cities." Grafton sits on the Park River, a tributary to the Red River of the famed Red River Valley of song.

She grew up just a block from the Park River, high enough not to be flooded, most of the time. During the summer months, which could be in the 100s, she spent a lot of time hanging out

Figure 40. Cindy Roberts.

at the river or at the local pool. During winter, which could be 80 degrees below zero with wind chill and lots of snow, Cindy skied behind a car and down bunny slopes, and built snow forts. As a child she liked riding horses and nursing injured or abandoned animals. She did not like sitting in church or doing homework, and she didn't like science, she said, "likely due to a dislike of the instructor." Her family, which consisted of eleven brothers and sisters, did everything together when she was young. They went to the several surrounding lakes, had picnics/BBQs, took drives in the country, watched movies and went to church together.

Cindy took general education courses at the University of North Dakota during 1977 and 1978, and then enlisted in the Air Force from 1978 through 1982. Following her military service, she earned an Animal Health Technician degree and other elective credits in wildlife conservation at Yuba College in California. From 1984 through 1990, she worked on her Masters Degree in wildlife conservation with a minor in chemistry at Sacramento State University, California, while working almost full time. She also had opportunities to take related courses during military service and in the course of her subsequent jobs with the Forest Service. She was attracted to her profession because she wanted to work in an area where she was able to help wildlife. What she loves about her job now is being able to make a difference in the quantity and quality of habitats for animals that are dependent on the quality and quantity of national forest lands.

After her military service where she had worked in avionics and electronics, Cindy followed her true heart and worked for a few years at the Yuba Sutter Veterinary Hospital as an animal health technician.

Next she was a student intern environmental services biologist at McClellan Air Force Base. There she managed habitat for pheasants and burrowing owls and provided input on sites eligible for State Historic Preservation.

From 1987–1990, she worked for the California Department of Fish and Game as a California spotted owl biologist,

Figure 41. Cindy (left) with coworker fisheries biologist Tina — Bear Ranch Creek, Plumas National Forest.

conducting surveys for California spotted owl in national forests in southern California, then went to the Shasta Trinity National Forest as their wildlife and aquatic biologist and botanist. Since 1995, she has been the wildlife and aquatic biologist on the Feather River Ranger District of the Plumas National Forest (Figures 41 and 42).

Day-to-day Cindy's work varies. She spends at least twenty percent of her time as a supervisor. This includes tasks such as hiring, terminating, approving time sheets, providing field equipment, work scheduling and training. She spends another thirty percent working on the budget, acting for her supervisor, planning the program of work for each year, representing the district or forest at meetings, facilitating meetings and she is responsible for the wildlife and aquatic program, including reporting and activities related to federal and state listed species and Forest Service sensitive species.

The above fifty percent of duties are the type of work avoided by those who opt to stay at lower levels in order to remain strictly or mainly in the outdoors. Still, district program administrators do enjoy time in the field.

The other fifty percent of Cindy's time is working on projects such as assessing impacts to wildlife from proposed vegetation management projects (timber sales, prescribed burns or other treatment of fuels), special use and minerals permits and forest-level projects such as relicensing power dams and off-road vehicle use on the forest. She also plans and executes wildlife and aquatic habitat restoration and enhancement projects.

Cindy did not deliberately seek out a non-traditional position. A lot of professional positions were male-dominated in 1991, when she started working for the Forest Service, and she did not think that the Forest Service would be any different.

When she first started with the California Fish and Game it was hard to be taken seriously. An upper management male told her that she should be "home barefoot and pregnant."

Figure 42. Cindy — Bear Ranch Creek landscape.

Another male employee told her that it was "my biological duty to have children." While taking out Forest Service employees for field and owl experience, she was told she was in a "man's job."

When she first started working with the Forest Service there was a lot of resentment towards women due to the consent decree, which had started a year earlier. Although she was not hired under the decree she was asked often if she was. Most men were friendly to her personally. However, many men make derogatory comments about women getting jobs they were not qualified for to satisfy the consent decree. Even older women in the agency and those that were office clerks were hostile to women in non-traditional positions. Cindy developed a thick skin, and things changed on the national forest she was on when folks realized, "I can do the job." She suffered a setback when she took a position on the Plumas National Forest, but it was not as bad as it had been on the Shasta-Trinity. She saw and heard of a lot of sexual harassment or hostile work environment but only a little was directed toward her.

Working as an intern for California Department of Fish & Game conducting surveys for the California spotted owl she got to know this species of special concern very well and saw more of California than folks born and raised there. That experience led to the next, which was working as a biologist in the Forest Service. She had a huge learning curve when she first started and she was not experienced with forest management (North Dakota is not very wooded except for riparian zones). Coming into a male dominated organization and a completely different profession from animal care, she had a lot of proving to do. She said, "I have learned so much in the past twenty years, it is incredible. I strongly believe I have made a difference for wildlife on national forest lands, which belong to all Americans."

~

Deborah Tibbetts, Archaeologist

Deborah Tibbetts (Figure 43) lived in or near Mill Valley, California, a suburb of the big city of San Francisco, most of her life.

As a child she was a tomboy, never went for the "girly stuff" until her teens. She always loved to run. Her family always camped out, and Deborah loved camping.

She developed a work ethic early, working summers beginning at the age of thirteen. She stayed with her aunt and uncle in Oregon, and all the kids there worked in the fields earning money for school clothes. She said that it was fun, and she made money.

She was an average student and really never liked school, and did not go to college until she was thirty-six. She has always been ambitious, and enterprising, though, unafraid of hard work and has been doggedly persistent when things got tough. And they did.

She was a fairly young single mother, raising her son on her own until he was fifteen years old, and then taking a troubled niece into their household. She met her second husband around that time. She said, "To say the household was crazy is a major understatement, but we made it through."

Since she didn't attend college when she was younger, most of her early jobs were the dead-end secretarial type. She did secretarial work until she was about twenty-five, then went into the field of keypunching, a now defunct pre-computer

Figure 43. Deborah Tibbetts hard at work.

technology. A big career change happened when she started working for a friend in the printing industry. She learned how to estimate jobs and finally did printing production. Eventually she got into printing sales in order to make more money. That led to a job as a printing broker for a private consultant in San Francisco, and then after a couple of years Deborah started her own business as a printing broker. She enjoyed the business immensely for a while, but she hated the cutthroat nature of it. She decided she wanted something else to do in her life.

Her turning point came when she and a girlfriend decided to go to college "on a whim." Deborah started at the Marin Junior College, because she flunked math and English in the college entrance exams. She recovered from that, and transferred to the prestigious University of California, Berkeley as a junior. She stumbled onto the area of study that became her passion when on another whim she took an introduction to anthropology class, and that was it; "the light bulb went off." She loved everything about anthropology.

In her senior year, at the age of forty, she got married, had a baby and graduated from college. At that time that she realized that she had to find a specialty in anthropology and get good at it; at least good enough to get a paying job. She took a year off from academia to raise her daughter, and eventually entered the master's program at California State University, Sonoma in the field of anthropology. It was difficult to choose her specialty. After a month on an excavation on the James River in Virginia on a 16th century plantation, she said, "I was smitten." She loved that you could research the historic record. That was when she chose historic archaeology as her profession.

After a year at Sonoma State her husband decided he didn't want to live in Marin any longer. They moved to Paradise, in the foothills of California, in 1989. Deborah transferred to Chico State University and completed her master's program in 1997.

After moving to Paradise, Deborah started work for the Feather River Ranger District of the Plumas National Forest as a temporary summer employee. Her supervisor, Kevin McCormick, recommended her for a permanent position at the supervisor's office in Quincy, working for the forest archaeologist as an assistant. She was still living in Paradise, and commuted nearly two hours one-way to Quincy for eight years.

Deborah was working full time, going to school full time and had a full time job at home with her family. Thankfully, her husband held down the fort while she worked. In 1998, when she finished a co-op program and was ready to graduate with her master's degree, the Plumas National Forest was downsizing and was unable to create a position for her. At that time she was working half time for the Lassen National Forest archaeologist and half time on the Plumas. Luckily, she was able to slide into the district archaeologist position on the Eagle Lake Ranger District on the Lassen.

She loved the Eagle Lake Ranger District, but said she felt "like a fish out of water, since archaeology there was centered on logging, ranching and prehistoric sites." Her expertise was Gold Rush Era archaeology. At this time, she was commuting five hours per day, to the Lassen's supervisor's office in Susanville, from Paradise. Just to make life more challenging, she and her husband divorced at that time. Deborah worked at the Eagle Lake Ranger District for three years and then got a one-year detail back on the Feather River Ranger District, finally easing her commute considerably. Fortunately, Kevin, her mentor and friend, was able to help her get a permanent position on the Feather River Ranger District where she had started, after certainly "paying her dues." She worked at the Feather River District until she retired in 2011.

Deborah never thought of archeology as a non-traditional job. She said, "I just knew I loved everything about it (Figure 44)." She started working for the Forest Service right after the women's consent decree was settled. At that time, the field of archaeology was also male-dominated, with few women in the field. Deborah was able to get extra financial help with college due to the underrepresentation of women in that field.

The special challenges she experienced were related to the ongoing lack of support and appreciation from the management staff towards employees, especially on the Feather River Ranger

Figure 44. Deborah (right) and fellow archeologist Robert Fred having a screening contest — she won.

District, which is separated geographically from the rest of the districts and the supervisor's office. The attitude, bordering on disrespect, was especially pronounced during her last ten years, and was the source of the worst employee working morale Deborah said she had ever experienced. Even so, Deborah cited one of the highlights of her working life as working for the Forest Service. She has observed that most everyone strives to do the right thing to meet the mission of the Forest Service, to serve the land and the people. Even with all the pressures, most love their jobs as she did.

With feeling, she said, "I have been gifted throughout my life with exceptional mentors, who have guided me in ways I would not have thought possible."

After retirement, Deborah volunteered part time at the Feather River Ranger District. She said, "I feel good about making a contribution to archaeology."

Chapter 4

Forest-Level Natural Resource Professionals

The Forest Service manages 154 national forests and twenty grasslands, each of which is divided into multiple ranger districts.

The administrative headquarters for each national forest or grassland is the Supervisor's Office, which oversees the management of its ranger districts, coordinates and balances the workload and budgets of the districts and provides technical expertise. Regional and Washington Office personnel that service forests may also be housed at a Supervisor's Office. Forest supervisors direct the work of district rangers and Supervisor's Office program managers, which together comprise forest leadership teams. Forests are the conduit for communications between the districts and the regional office. Forest-level employees perform upward reporting about forest activities to the regional offices. Regional offices disseminate budgets and direct the programs of the forest level, which the forests then distribute to the districts. Forest staff specialists work with district rangers and district-level staff to coordinate programs.

This chapter covers the stories of several women in forest-level positions, and their earlier jobs that led them to those positions.

Jodie Canfield, Wildlife Program Manager

An independent, colorful character from the start, Jodie (Figure 45) grew up in Great Falls, Montana and went to high school in Anchorage, Alaska. She left home at seventeen and spent the last semester of her senior year in Great Falls, living in a small house in the back of an old folks home. She graduated in the top ten of her 650-plus-student class, in spite of many shenanigans as a senior without parental supervision.

Figure 45. Jodie Canfield.

Jodie describes herself as a relatively solitary child from a family of three girls. Her mother worked at a fabric store, and her father was in the Air National Guard. She liked camping, loved school and liked playing in the neighborhood with other kids. She was skinny and disliked eating and "anything to do with doctors!" She liked all subjects and was equally interested in the "left brain" subjects like science and math, as well as art, English and even home economics.

Jodie's family was not especially outdoorsy. Her father hunted and tried to take her, as the most likely girl child, but she was not that interested in killing things. She was more interested in what made them do the things they did. Her father lost interest in her as a hunter, and she got more interested in small creatures that people did not hunt.

Jodie started college in general studies. She remembers looking through the course catalog at Montana State University, seeing the course work requirements for fish and wildlife and "resonating" with that, and so deciding to declare that her major. She said, "It was an uphill battle for women in those days; there were only a handful of us. I loved being in the field and watching animals; it was very 'Zen' for me."

Jodie had a lot of great field jobs in her early career that kept her addicted to wildlife biology. She surveyed salmon streams, observed prairie falcons in relation to dam reconstruction, radio-tracked sage grouse on a nuclear facility, measured elk hiding cover on the Lolo National Forest and finally, in graduate school, examined the effects of the Bonneville power line on elk habitat use.

She loves problem solving, in relationship to wildlife and habitat relationships. She was the wildlife program manager for two adjoining forests, and enjoyed solving problems at many scales in that position. She was particularly interested in some very key vegetation communities, namely aspen and white bark pine, and she was involved in regional and local efforts to restore those wildlife-rich communities.

Jodie counts childbirth as her number one, "over the top" extraordinary experience. She spent a year as a full-time mom when her child was a senior in high school. She has been "happily divorced" for thirteen years and this was the first time she had full-time custody of her son. They are both Aries, and "locked horns" a lot. Jody said, "He is amazing, and amazingly frustrating."

In terms of extraordinary wildlife experiences, Jodie once saw a great gray owl take a goshawk chick from a nest, just as lightning struck the nest tree. On a different level, she was awarded the first ever "individual achievement award" from the Rocky Mountain Elk Foundation, and she got to be part of a district team under an amazing ranger where she said, "We kicked butt and took names for about seven years. We did some great stuff for the 'ground,' back in the days before excessive appeals and litigation."

Jodie did not deliberately turn to a non-traditional career. She worked for Montana Fish, Wildlife and Parks for several years before applying for a Forest Service job. The Forest Service was relatively liberal towards women compared to the state.

She said, "I did what I was passionate about; not influenced by tradition. I think I have always been driven by my passion and that overcame a lot of challenges that I might have faced, if I had been more timid. It was a challenge to balance life and work once I had a child, and even when my child was a senior in high school."

Due to organizational issues, Jodie spent five years of her Forest Service career in a job that was very demanding, but not her passion, including managing the timber program for the forest. When she got back into wildlife she had a renewed sense of passion, but she missed the role of supervising and team building and working with people. Her passion is for the land, "the public land that we have the privilege of managing. The bureaucracy is what it is and will always bring frustration. But the landscapes of the forest are inspiring, as are the wildlife species that call them home!"

∽

Stephanie Connolly, Soil Scientist

Stephanie Connolly (Figure 46) was raised in Appalachia. She spent her summers on her grandparents' family farm where she fished, drove tractors with her grandpa, helped out around the farmhouse and gardened with her mom. During school, she played outside with her friends, built forts, snow sledded during winter, ice-skated and whatever else they could do to pass time outdoors. She loved the freedom to run and play.

She disliked rules, restrictions and boundaries. She was raised Catholic in a very large strict Catholic family, from both her Irish dad's side and German mom's side. She said, "I still maintain the stereotypical guilt today but do not practice my religion."

She loved school, science and math being her favorites. She did her first dissection when she was five. She participated in

Figure 46. Stephanie Connolly — front lawn, Monongahela National Forest.

social studies fairs and science fairs, even doing more than one project for the just-in-case scenario.

Until she was about fifteen, she thought she would be a marine biologist, but changed her mind when she realized she would have to live by the ocean. She could not be without her mountains. Years later she learned that she could be by both mountains and the ocean but at that point she was more in love with the soil than anything else.

She was drawn to the soil at a very young age. She had her first garden at three with a picture to prove it (Figure 47).

She even did her own weeding. Working with her grandpa and uncles in the fields also felt so right as a little girl. At the end of the day, she could see her work done when she looked back over the field, no matter whether they were plowing, bailing hay, or planting. Later in college, the mine reclamation industry was her focus. "Talk about a landscape to paint a portrait of new soil on (Figure 48). That was amazing to me," she declared.

Figure 47. Stephanie's fond memory — gardening with Dad.

Figure 48. Stephanie in fall foliage, Mower Tract Restoration Project, Monongahela National Forest.

Figure 49. Stephanie (left) and Interns at NE Regional Soil Survey Conference, Whiteface Mountain, New York.

She said soil science is, "my soul's passion but not my life's passion." She believes soil forms the fundamental building blocks of society, that, along with water, it sustains the world. Her interest and passion are evident in her statement:

> To know something about soil is to know something about our own origin of life. Soil is so different depending on the characteristics of the point you stand on the surface of the earth. What I love most is that soil seems to be a secret and when you share that secret with someone new . . . you can see them start to change their very thought processes as you talk to them and they begin to understand just how important the soil is (Figure 49).

Her next favorite thing is to hear someone who is not a soil scientist cite something about the soil and how important it is to someone else.

Stephanie services the entire forest and acts as a liaison/consultant to universities and other government entities in the region, United States and internationally (Figure 50).

She may handle many different topics in a day and deal with multiple resources as she provides soils information to the folks on forest and off to help with their projects.

Figure 50. Stephanie and Dr. James Thompson field review the new soil series Gaudineer.

Figure 51. Stephanie (right) and interns digging soil pit.

Her favorite days are when she gets to head straight to the field to dig soil pits with a crew (Figure 51). She said, "Soil pit digging is by far one of the most rigorous activities in the USFS workplace — just ask all the tough guys that have dug pits for me over the years how they felt the next day! Ha-ha!" By midday she finds herself inches from the soil profile, describing in detail the way the skin of the earth looks ". . . and how beautiful she truly is."

Stephanie did not deliberately enter a non-traditional profession. Men supported her early in her life to work outside. But even when she told her grandpa that she wanted the farm, he said she was too smart to farm and needed to be a doctor or lawyer. She says, "So I showed him . . . I went and got an agricultural degree!" Then in college, she had a professor that would tell her she could do whatever she wanted in life and sent her on active mine sites to collect water samples, of which she said, "Boy was I a sight in my tank tops, cutoff blue jean shorts and mining boots."

By the time she was in graduate school, she could dig pits and carry huge bags of soil, no problem, and there were more females in her progressive graduate school department than there were males. She does have some great experiences to share about being in a male-dominated field. She has been told that she was taking jobs from a man who would need to feed his family. Her response was, "That was too bad this man wasn't smart enough then to get off his ass and get the degree and fill out the application for the job." She has been asked why she was not at home having kids. She responded that she really didn't like kids or even people in general. She has been led to believe that her safety was not important. Her response was to file a request for a satellite phone from "Mother USDA," at a time when they still cost thousands of dollars or she would file a suit against her supervisors. She said she has "been hit on more times by men than you could shake a stick at . . . and so on and so on . . . and, oh . . . I have had my boss (in the far past and not in the USFS) literally turn around and pee within feet of me while still talking to me. It has been a wild ride at

Figure 52. Stephanie and Pathways Intern Adrienne Nottingham in Budapest, Hungary.

times. I would not trade it for the world because it makes me who I am today and I have the understanding of why some things are the way they are out there and how to get around the obstacles."

Stephanie said she often finds that the sexual differences between males and females in the office are not the problem especially if, "I learned to be just as vulgar as the men." She found religious differences to be more of a problem. Religion always seeped into the office environment. This is often most difficult because it seems to her to really dictate how people judge others and how they work with others. She considers that educating herself was always her best defense along with "a move in the near future to a place that seemed more tolerant of differences."

The desire for stability, maintaining integrity and credibility drive Stephanie now. Being able to consistently do the things that she says she is going to do is important to her. This type of accomplishment pushes her to do even more.

For example, in 2015, Stephanie and Adrienne Nottingham, a Pathways soils intern on Stephanie's forest, were members of Team USA in the World Championship for Collegiate Soil Judging, held in conjunction with the International Field Course and Soil Judging Contest in Gödöllő, Hungary (Figure 52).

Team USA comprised the top four winners from the national competition held at the University of Arkansas, Monticello. Stephanie served as assistant coach for the world team, and Team USA won the 2015 World Championship for Collegiate Soil Judging.

Stephanie loves being a high level efficient operator with energy and vision to go even further if she pushes herself. Her passion and greatest goal is being true to her self, "living everyday with no regrets and not being afraid of dying in the next moment."

∽

Patti Fenner, Forest Noxious Weed Program Manager

Patti (Figure 53) was born in Baltimore, Maryland, and raised in the Phoenix, Arizona area where her family moved, in an un-air-conditioned station wagon, when she was two.

Her family was not an outdoors family. Her dad, an electronics engineer who worked at Motorola, did like to fish, even made his own flies and fishing rods. But he rarely took Patti and her sister with him. Patti said, "He learned early to take only one of us at a time, or he'd have to give up his own fishing time to the endless chore of untying knots." Patti's mother was a librarian, but liked to garden and always had wonderful flowers in the front garden. Patti still remembers the smell of the nasturtium bouquets she would bundle up for her to take to her first-grade teacher.

Patti had an affinity for plants. She spent lots of time playing alone, and fooled with the lantana, dichondra and orange trees in her yard. She said, "I didn't know orange trees weren't made for climbing, and always had hideouts as high as you

Figure 53. Patti Fenner rides Sugar into Chalk Mountain Grazing Allotment in the Sonoran Desert.

could climb in one of those squatty branchy trees. I even taught my dog to climb the best tree."

She discovered the more formal field of botany her first year in college, when she spent the summer in San Diego, California. She found it a wonderful place, where plants that are only houseplants in Arizona grow in profusion along freeways and in wild places.

Patti found a home with the Forest Service, starting out in the job of recreation technician, where she spent many hours with friends she still has to this day, picking up garbage and cleaning outhouses. She felt she had finally found people like her, and they always had a great time, no matter how nasty the work. Some of those friends have gone on to lifelong careers in administration, fire and range management in the Forest Service, or high-level fire positions in the Park Service. Some run their own businesses now. Some are not around anymore, and Patti feels lucky to have known them in their heyday.

They worked for a redneck ranger, who believed men's hair should be cut above their collars. Not knowing how to treat women, but careful not to discriminate, he required them to wear their hair above their collars too. They all screwed their hair up into little buns, which sometimes lasted through the entire day. There was no women's uniform — they all wore men's Forest Service shirts that were too big, with blue jeans. Patti remembers going on fires, and having the rivets in her jeans getting pretty hot.

They all had basic fire training, and when the fire bell rang, the recreation folks jumped into their trucks to go to fires with the fire folks. Fires were pretty exciting, even though that first year most of them seemed to be mostly out by the time they got there. There were some large ones, and Patti learned to respect the power of a chaparral fire.

Patti was first hired as a permanent employee on the Tonto National Forest as the result of a lawsuit. These days, when environmental groups sue the Forest Service, there are no extra funds to deal with the extra work. Back in 1980, though, several additional jobs were created on the Tonto to deal with an Audubon Society lawsuit. The lawsuit claimed that the Tonto was not managing livestock grazing to a standard that would allow cottonwood trees to grow up and provide nesting habitat for the endangered bald eagle along the Salt and Verde Rivers. At that time, there was grazing during the summer, sometimes yearlong, on most all of the grazing allotments along the Verde River. Those allotments were large — 60,000 acres or more — and remote. Much of the country lay in the Mazatzal Wilderness, inaccessible by vehicle. The Cave Creek Ranger District hired Patti and a new biologist to help with doing "Level IV Range Analysis," creation of new grazing plans that would allow riparian vegetation along the river to recover and negotiation of reduced stocking rates on all of the allotments.

Their boss rarely went out with them, but Patti spent all her time with the range technician, Buck McKinney. He taught her how to pack a mule so the pack would not fall off even if the mule buggered. "Well, most of the time," Patti chuckled.

They packed for ten days at a time — they called it ten and four, because it was supposed to be ten eight-hour days, with four days off. But the summer days are much longer than eight hours, and when you are out there, "you may as well be working as sitting around swatting flies."

Their home away from home, in the middle of the Red Creek Allotment, was the Tangle Creek Cabin. It was homey, with a screened-in front porch where you could put the bunk beds for comfortable sleeping in the summer. It had a well, running water, a sink, wood stove and a propane stove and refrigerator, with a water heater and metal shower stall right inside the cabin. A corral was a short walk past the outhouse and barn.

Patti learned to read range transects, called Parker-3-step Clusters, named after Doug Parker, who came up with the monitoring method in the 1950s. She remarked, "That was always fun, especially finding the short pieces of angle iron stake pounded into the ground ten to twenty years earlier, that marked the beginning and ending points of each of the three transects." This was pre-GPS, and they spent a lot of time dreaming up ideas for how to find the stakes more easily. Ideas included metal detectors and specially trained hounds. They never imagined using satellites in space. They rode every ridge, every valley, mapping the thousands of acres of range and classifying its condition.

They spent most of those first years with a rancher named Glen Jennings, who had grown up in Tonto Basin. His mother was the schoolteacher there in a one-room schoolhouse. Glen was a gentle man, but tough. He did not have a lot of formal schooling, but he knew the names of all the plants on his ranch, and he knew the place like the back of his hand. Patti said he was smart, and fun to ride with and talk to. She said he always said the time you spent reading clusters was in reverse proportion to the number of people you had helping, due to all the discussion that happened.

In those days, the ranger district maintained a herd of about eleven horses and mules. Part of the job was caring for the animals, which Patti enjoyed. She said, "We used the animals hard, but we took very good care of them." She learned how to give shots to the animals, to save a little money on vet bills. Cave Creek had a big parade every spring, and the Cave Creek Ranger District was a part of it. One year they painted their mule's hooves bright red, applied lipstick to her whiskered lips and attached silk flowers all over her bridle. Patti said, "Marty was really something special. She was the ugliest mule you've ever seen, but in a tight spot you were always glad you were on her. She didn't freak out at rattlesnakes, but would calmly walk *way* around them."

Patti spent later years on the Sears-Club/Chalk Mountain Allotment, where they had to drive with a horse trailer for a few hours, unload and pack up and ride five more hours to get to Club Cabin. Club Cabin is a wilderness line shack built in 1902, where they based their camp. Many times they arrived there after dark, and depended on experienced horses finding their way to the cabin, "the way horses can do, even in the dark."

In 1990, Patti had her only child, a son. She remarked, "I am still convinced that he was someplace watching me and my husband, and he's the one who decided to come live with us. His personality has unfolded like a beautiful young plant, and I have always marveled at how we could have created such an incredible human being."

When their son, Eric was in grade school, she and her husband felt lucky that the Forest Service let them modify their work schedules so that each of them worked a four-day workweek of ten-hour days. They overlapped their schedules by one day, and on that day Eric would go to the field with one of them. The other days one of them was home with him. There aren't too many workplaces that make allowances like that for families.

In the late 1990s, after another lawsuit, things changed again. They had to monitor riparian vegetation along the Verde River three times a year, with no extra funding this time. Since this country was very remote, it involved learning how to use a kayak for transportation. Luckily, the Verde River is a somewhat forgiving river, and "didn't kill me while I was learning to maneuver the inflatable kayaks we used." Even though the forest had hired a riparian monitoring team by that time, Patti decided it was part of her job as the district range staff to conduct that monitoring, and she did not get a lot of argument about that from the riparian specialist or monitoring team. She loved those years of Verde River weeklong trips, and says she will never forget all that she learned from observations and monitoring.

Patti enlisted the help of an Arizona State University range professor, and they attached metal tags to over 800 cottonwood and willow trees. They tracked the utilization and growth of those trees over three years, to be able to objectively study the effects of livestock grazing along the Verde River. What they found was that other factors, such as soils, crowding and other characteristics of the specific site, like vulnerability to removal by flooding, affected the trees much more than winter seasonal grazing. They also learned that politics could override scientific studies, when all grazing was removed from the Verde during creation of the Verde Wild and Scenic River Management Plan.

Another major change came in 2003. Patti's ranger called her one evening and asked if she would like the job of forest noxious weed program manager. This would be a totally new job created on the forest, and she would get to initiate and run the new program. He told her she had to let him know that night. She stated, "How could I resist? After doing this job for eight years now, I feel that I've been sidetracked for about twenty years, and have finally found the job I was meant to do." She said she did wonder at first how she would get over her fear of talking in front of audiences, but even that worked out all right. She found, "If you are enthusiastic about the topic, it's all good."

Those days Patti surveyed for and mapped weeds, worked with partners outside the Forest Service and did weed control projects with grants and volunteers, and with other Forest

Service people. She could have retired years before she finally did. When people asked her when she was going to retire, she told them, "I'm in no hurry — I just got the job I've always wanted, I'm productive and I'm having a great time."

A few years ago, Patti did a presentation for one of the Phoenix fly fishing clubs, about how invasive plants can affect their sport. Afterwards, an old man walked up to talk. It was old Doc Miller; the veterinarian Patti's father took their family dogs to for many years, and a friend he used to go fishing with. Two local fly-fishing organizations have now adopted weed control along Canyon Creek, a blue-ribbon trout fishery and one of her father's favorite fishing spots. Patti said, "They come out every year and we work together to control bull thistle. I think my father would approve."

Patti did retire in 2014. Well, kind of — in 2015, she formed the non-profit Friends of the Tonto, and now serves as the organization's executive director. Her husband, an archaeologist, also "retired" from the Forest Service in 2015, and is on the Friends' board and chairs their cultural resource committee. Friends of the Tonto enlists and leads volunteer groups to help the Tonto National Forest accomplish its mission of natural and cultural resource management.

Nadine Pollock, Ecosystems Staff Officer

Nadine (Figure 54) was born and raised on the Allegheny National Forest, started her career there and lived there until her first transfer with the Forest Service. She kept her home and was able to transfer back to that forest a few years ago, and will retire there.

Nadine's family did not camp or hike, but did scenic driving, her dad discussing the historic places on the forest, like the CCC camps that were sprinkled around the area, including the one he was part of. They also did a lot of watching wildlife, picking berries and picking wild mushrooms. Nadine remembers those as wonderful experiences and some of the best childhood memories she has. She thinks the reason they

Figure 54. Nadine Pollock — Festival of Wood at Grey Towers.

did not go camping was because growing up and living right in the middle of the national forest meant that they could be out playing all day and then just come home in the evening!

When Nadine graduated from high school, she did not know what she wanted to do so she did not attend college right away. She wanted to work for the Forest Service. She thinks that interest came from growing up in a small Forest Service town and attending school with kids whose parents worked for the agency. She said, "Of course, there was always Smokey Bear — I knew exactly where the coloring pages were in the Sheffield district office-on the wall rack right behind the entrance door!"

With only a high school education, the jobs she was qualified for were limited, and it took a long time for vacancies to occur. Finally, she was hired into the Office Services group in the Allegheny National Forest Supervisor's Office. That job was her foot in the door in 1988. Once she was hired and saw all of the opportunities before her, she decided to start college, at the age of thirty.

She was interested in public and legislative affairs. Her first round of college resulted in an Associate of Science in business administration. Then she transferred from the Allegheny to the George Washington National Forest and finished a Bachelor of Science in public administration with a concentration in political science. She worked full time and some semesters attended class ¾ time. It was a rough few years. At the same time, she felt fortunate to be working as a public affairs specialist. To sum it up, she said, "My desire for a good stable job in the clerical arena, with no intention to ever transfer from the Allegheny, landed me where I am today. Once I got my foot in the door I fell in love with the agency and all the opportunities it offered. My work duty stations included the Allegheny, the George Washington/Jefferson, the Washington Office National Fire Plan and Fire and Aviation Management and the Monongahela National Forest."

Nadine's biggest challenge was not having a natural resource education, but wanting to be in positions that required it. She had no energy to go back and get a resource degree. However, the work she did as a public affairs specialist gave her knowledge she needed in the resources to move into positions that she were once only for foresters and "ologists." She said, "I am thankful that the people I worked with who were foresters and 'ologists' were great mentors and teachers and educated me well. All of this brought me to the last couple of positions I've been fortunate to have."

What drives her now is doing the right thing for the natural and human resources entrusted to her care. She said, "I feel very fortunate to have had the career I've had to this point and want to share with people who are either just starting out or are in mid-career and could benefit from me sharing my experiences. I want to help people be successful. It's very rewarding to me to be able to mentor someone or submit them for awards for their accomplishments for all the good they do."

Nadine is proud of her accomplishment of completing college the way she did. She said, "It took a lot of time and energy to attend college and keep a full time job and do it all to a high degree of quality. There were days and times I didn't think I could do it all. On those days and times, my mentors stepped in and coached me through the rough times and renewed my strength to carry on. I'm so thankful to them!"

She counts as an accomplishment being selected for a staff officer position for natural resources and ecosystems program areas, which are generally held by traditional foresters and "ologists." She feels lucky that her supervisor was willing to give her a chance to show that a person without a traditional resource education con do the job well.

The Forest Service can be equally thankful for Nadine's interest and energy for doing an excellent job. Here is what some of her day-to-day activities look like:

She tries to start work between 6:30 and 7:00 a.m. for a little quiet time to catch up on messages. Most days have very little alone time as she is constantly interacting with members of her teams on current projects or the "crisis of the day."

That could be anything from having to change plans of a helicopter seismic operation on one of the ranger districts, to reviewing a burn plan for a timber sale project, or giving input on the new design of a campground to get the sites out of the flood plain.

There are interactions with external scientists, sitting in on interdisciplinary teams, or reviewing work done under agreements with cooperators and partners. She puts in a lot of time overseeing a budget of around five million dollars.

Nadine works non-stop all day. She tries to keep some sort of schedule. She said, "It's rare that I have a couple of hours free to do 'administrivia' during the workday. Somehow it gets done and as near as I can figure, it's just piecemeal! Most of the time, the associated stress with this workload is good stress. We have some people around who can give you the bad kind of stress too and be relentless about it."

Nadine says that getting to the field preserves her sanity. Plus, when the specialists have issues with the projects, she needs to see them on the ground in order to be able to help them in some way, either for their program or to find some middle-ground compromise between multiple program areas. She wants to be in the field more, to keep learning the programs she has taken on in the ecosystems area.

On her way to her current position, Nadine worked in a variety of increasingly responsible jobs. When she started on the Allegheny National Forest, she was trained in the office services group and typing pool. In a short six months, she was selected to be the lead receptionist and manage the reception area. She transferred from the Allegheny in 1992, to the George Washington National Forest as a visual information specialist where she was responsible for all of their publications, some Freedom of Information Act (FOIA) request responses and some entry-level public affairs work. After a couple of years, and through the merger of the George Washington with the Jefferson forests, she was reassigned as a public affairs specialist with full public affairs officer (PAO) duties for the north zone of the forest. She then did a long-term detail back on the Allegheny as their public affairs specialist for a year, focusing on several big environmental impact statements. From there she went to the Washington Office Fire and Aviation Management on another yearlong detail. In January 2003, she was selected for a promotional move to the National Fire Plan (NFP), where she was assigned work in communications and information management for the plan. She worked a lot with Department of the Interior (DOI) people as they had a mirror national fire plan organization. Then in about 2005, the NFP staff was absorbed back into the original fire and aviation management group and Nadine was their communications and legislative liaison. She said, "It's a nice was of saying I did PAO work for them!" During most of her WO time, she worked closely with the national fire information desk either scheduling detailers to come in, covering for someone who cancelled, or providing information they needed on NFP efforts. She left the WO in September 2007, and went to the Monongahela National Forest as natural resources staff officer where she was responsible for oil, gas and minerals, recreation, heritage, wilderness, fire, timber and silviculture. During that time she was also assigned lands. In March 2012, the leadership of the ecosystems group (ecologists, hydrologists, fisheries, wildlife, botany and partnerships) was added. Of all this, she said, "WHEW — I can retire in 6½ years."

None of her jobs required Nadine to be field going, but she made them so, especially when she was a PAO. Media was something she really enjoyed and was fairly good at. She did a lot of photojournalism, as the papers in the towns where she worked were hungry for a complete story and rarely had enough reporters to do what she did. She said, "So, I spent at least a day every couple of weeks going out taking photos and gathering information for my next story. Maybe someday, I'll do that in my spare time." She also did a lot of fire information work, but no longer has information officer qualifications on her red card (documentation of one's fire duty qualifications). She quipped, "The best I can do right now is be the lunch truck driver, which is okay because the fire fighters need to eat!"

∼

Deb Sholly, Minerals Management Staff

When Deb (Figure 55) was born in 1951, her father was chief ranger of Big Bend National Park in Texas. From there they moved to Shenandoah National Park in Virginia, and then transferred to Badlands National Park in South Dakota, where her father was superintendent of the park. He died there when Deb was eight. Because the family was in government housing, her newly widowed mother was given thirty days to vacate. They moved to California and lived with Deb's grandparents. Her mom married a cattle rancher when Deb was twelve, so off they moved to New Mexico.

Deb went to college for a year, partied too much and flunked out. She moved to Grand Canyon National Park in Arizona at nineteen, and got a job with the concessioner there, operating a switchboard at the Bright Angel Lodge (Figure 56).

Figure 55. Deb Sholly — born to run.

She later worked for the local hospital, and then got a permanent job as a dispatcher for the National Park Service. She met her husband there and married on the rim of the Grand Canyon. After three years they moved to California. Deb got pregnant, and during her pregnancy, they went to Washington D.C. for a three-month assignment. Her son was born three months early, in Alexandria, Virginia, and so they traveled coast-to-coast with a five-pound baby, to get back home to California. Deb worked as a temporary employee doing nature walks and collecting campground fees. They transferred to Canyonlands National Park in southeast Utah when her son was a year old (Figure 57). When he was four, Deb and her husband separated. She said, "This is when my life truly began."

She became a professional river guide on the Colorado River in 1978, which she did for nine years. She also worked for a year for the State of Utah as a river ranger for one season.

Figure 56. Deb's early days at Grand Canyon.

Figure 57. Deb hiking in Utah.

At the age of thirty-two, she returned to college in Salt Lake City. She went to school in the winters and ran the river out of Moab, Utah during the summer. She moved to Park City and got into the ski culture, working in the winters at a retail/rental shop at the base of the ski area, skiing over 100 days per year. She commented, "This really cut back my school time!" When she finally graduated with a degree in anthropology, she moved to Missoula, Montana to attend graduate school, but decided to study something different, so she switched to recreation management and earned a Master of Science with an emphasis on resource management.

Upon her graduation from college, Deb wanted to work for the Forest Service. She knew nothing about the application procedure, so she simply wrote an introduction letter, made twenty copies and sent them out directly to "Forest Supervisor" of the different forests where she was willing to work. She said, "I should have done my research, but I've always had a faith that things would turn out, somehow, if I was assertive enough and just 'put out there' what I wanted!" In this day of such specific application procedures, it is amazing that her letter ever made it through the right channels, but within one week, she received a call from the archaeologist on the Tongass National Forest in Alaska, who offered her a six-month long seasonal job as an archaeologist. Not even realizing at the time what a huge gift this was, she thought, "Sure, why not?" Her son was seventeen at the time, and after some discussion, he was game to change schools for the last three months of his junior year, in the interest of having a new adventure that none of his classmates would share. Deb parked her car for the summer; they packed their backpacks and flew to Seattle to make their way to Bellingham, Washington, where they got on the ferry to Ketchikan. Deb was too poor to purchase a stateroom, so she and her son pitched their tent on the deck of the ferry for the three-day trip. She really didn't have much money, and her son was a typical teenager, always hungry. Ferry food was expensive, so they ate lightly during the trip. One of Deb's fondest memories is arriving in Ketchikan.

One of the other archaeologists picked them up and they went out to breakfast. Her son wolfed his breakfast down, then ordered a second breakfast and completely devoured that one. Deb said, "The poor kid was just starving and it was a joy to see him finally satisfied, as we began our adventure in this new place."

After living in dry Utah for many years, going to Ketchikan, with 161 inches of rain a year was an interesting surprise. Deb owned a nice little Gore-Tex raincoat, which everyone laughed at. She was taken in hand by a fellow archaeologist and taken to the hardware store, where she was ordered to purchase a set of large, heavy rubber-like bib overalls, a bright yellow jacket with a hood and knee-high rubber boots, called "Ketchikan Sneakers." She said, "I can't hike in these ... I HAVE hiking boots!" Again, she was laughed at and told to buy the damn boots. By the second day, she was out in her red rubber boots, up to her knees in water and most grateful for the dry feet (Figure 58).

They went out to the field every Monday by floatplane, and flew back to town every Friday. Back then, the Forest Service had functioning field camps, usually floating barges, and they would go out to the field for the day and return to the camp at night Figure 59). To Deb, it was no hardship: there was a live-in cook and a maintenance man to keep the generators,

Figure 58. Deb suited up for a chilly day at work on the Tongass National Forest.

Figure 59. Deb at work in remote location, Cholmondeley, Alaska.

vehicles and boats up and running, so even though they were out all day hiking steep, muddy hills in the rain and wind to clear timber units, going back to camp was a joy. To Deb, who loved the outdoors, being "out at camp" was an adventure, somewhat of a vacation. On weekends, she learned how to fish for halibut, salmon and crab (Figure 60).

She said it was a beautiful summer, and she learned so much. Deb recalled two things that people shared with her that have stuck with her: "Yes, it rains here all the time, but when the sun comes out, it's the most beautiful place in the world." She found this to be true and has used it during difficult times. She said, "it is so indicative of life…when the hard times come, I have faith in knowing there will be something good coming around the corner ahead." Another time she was complaining about wanting to go kayaking, but it was pouring rain…again. Her boss told her, "Put on your rain gear and just go outside and do it. If you let the rain dictate your activities here, you'll never do anything!" She learned right then to "suck it up and just go… *I* am in control of what I do (Figure 61)."

Deb returned to Utah to finish her last year at the ski area in Park City, then in the spring, packed up her truck and moved to Alaska without a job. She stayed with a friend that worked for the Forest Service, and had intended only to spend her summer

Figure 60. Deb sockeye fishing — Alaska.

Figure 61. Deb kayaking — Ketchikan Creek.

sea kayaking with a new boat she had purchased. However, an offer for another job with the Forest Service materialized, so she took it and spent three years as a seasonal technician in the field, first working in soils, in the office, updating the GIS soils layer, "which was a really bad job," then for the area ecologist, doing field studies on an EPA buffer stability study, which "was a great job, because I spent every Monday through Friday out in the woods." She said the worst part about that job was the actual work — assessing buffers that had been purposefully left, next to timber units. This entailed hiking daily through a clear-cut, and measuring blow-down — no easy or fun task. She said, "You just can't possibly understand the size of root wad of a large hemlock or Sitka spruce until you attempt to climb over one that is more than ten feet high." It was truly a physically demanding job — Deb said she would not recommend a hike through a clear-cut to anyone.

In 1995, after starving for three unemployed winters in Alaska, it appeared that Deb was not going to get a permanent job, so she went to California and attended massage school. After becoming certified in three types of massage, she surprisingly received an offer for a permanent job on the Tongass National Forest in December, as a biologist working in lands and minerals and special uses. She cited that as, "Yet another example of the universe coming through when it is time." She moved to Prince of Wales Island and spent six years in the field, doing all her fieldwork from helicopters, floatplanes and boats. (Figure 62).

Her favorite memories are of when she would take off for a week alone in her boat, and travel around the various islands doing inspections on mining claims or special use permits (Figure 63), spending nights in old Forest Service cabins and staying in touch only during the day by radio to the dispatcher in Ketchikan. She said, "Boating around by myself through Southeast Alaska's outside straits in an 18-foot Whaler were probably the best days of my life, bar none."

In 2001, after nine years in Alaska, Deb transferred to West Virginia as the minerals specialist on the Monongahela National

Figure 62. Deb piloting a friend's boat, Vindicator, on the Tongass National Forest.

Figure 63. Deb inspecting gas tower special-use permit on the Tongass.

Forest, a field position requiring her time on gas well and pipeline construction and inspection.

Compared to her time in Alaska, her job in West Virginia was a cakewalk. She said, "There were roads, and I could actually drive to my job sites!" West Virginia was somewhat of a shock to her, socially, however. Although the people were friendly, it was still a fairly closed society. Numerous locals work for the Forest Service there, and were a different breed of people than the Forest Service people Deb had come to know. She found it is difficult to make friends there, and for a single person, it was a lonely place to be. She was grateful for her upbringing and her ability to enjoy alone time, because even after six years there, she left West Virginia with no West Virginia friends. She said, "It is a beautiful state, however, and even after being gone from there now for years, I miss the sweet smell of the springtime and the gorgeous fall colors of autumn."

Deb's mother used to tell her that as a child, when in a group, she would take some toys into a corner to play alone, that she never wanted to play with a group of kids. Of that she said, "I often wonder if there is something in my personality that led me to ultimately search out a job away from people, or whether it was my upbringing in the National Parks system. I was certainly at my happiest when I was alone and out, roaming through Mother Nature, especially if I was in the woods or in a boat of some kind."

When she was a child, Deb's father took a month off every summer, and they took a road trip and camped every night. In the 1950s, rangers didn't make much money, and they could not afford motel rooms for their family of four. Deb's first camping trip was when she was six months old, so she has always felt most at ease outdoors. At Girl Scout camp, at age ten, Deb set up her camp alone and slept in the woods away from the others. She said she wasn't trying to get away. She just did not know anyone and was intimidated by all the other groups of little girls that were friends and all seemed to know each other. However, she said that looking back, she believes this solo camping trip may have built up some early comfort and confidence in being alone in the outdoors.

As a river guide, Deb took folks out for five days at a time and camped along the river, cooking for them and being their interpretive nature guide. But she says she much prefers the outdoors without the people. She is not a group person, and is just more comfortable alone or on a one-to-one basis.

Deb said she is not sure she turned to the non-traditional as a choice. She said, "I think it not only fit my personality, but moving around so much as a child built in an odd sense of both loneliness and self-confidence. We never lived anywhere long enough to make friends, so I learned to take care of myself and enjoy being alone."

In terms of being in a male-dominated field, she preferred it, but it did not come without a struggle. She said that during her river-running days, it always felt like the old adage, "I must work twice as hard to be half as good." Additionally, it has always been difficult for her to be around the dramas that seem so elemental to some women. She is the youngest in her family, having two older brothers, thus has always been comfortable with men. She remembers being very jealous of her older brothers when they got to do exciting things and she was not allowed to, because it was "too tough" or because she was a girl. She is sure that those feelings contributed to a sense of "I can too!"

Besides, she always thought that men had more interesting and exciting jobs. Her stepfather taught her how to shoot a gun, and she can remember her first job in the National Park Service; she was nineteen years old, and she worked as a dispatcher for the law enforcement division. They took her out shooting with them when they had to qualify with a gun, and were surprised when she not only qualified, but also out-shot a number of them! While it did not make her feel better than them, it taught her a very important lesson: she could compete as long as she had the skills.

That came with challenges. As a summer seasonal with the Forest Service one year in Utah, Deb was handed an enormous chain saw, and with some very basic training, stuck out in the woods with a crew of men, cutting down trees that had been affected by the parasite mistletoe. It was rough, as she was the only woman on the crew, thus they had to provide her with her own room and bathroom in the field camp, which the men resented. This was in the late 1970s, and some guys on the crew resented her so much that they would not even speak to her. Although she was strong and in excellent condition, she lasted only a week before she had to confess that she was holding the crew back. She just could not physically keep up with them and move as quickly as they did. She had to go to the boss and ask him to reassign her. She said, "It was the only job I ever backed away from, but I never felt bad about it, because I knew I had given it my best."

As a river guide, in order to run the big boats through Cataract Canyon, one had to be able to lift and move two 98-pound motors. At the time, there were very few women running motors; most of them rowed rafts on the smaller sections of river. Deb began lifting weights until she could bench press 140 pounds so that she could lift the motors and prove that she could perform the job. She knew she had "made it" when on a trip with two other guys, they helped each other move their motors, but did not offer to help her. She lifted and loaded her own motors, and after a couple of years, she earned their respect, but she said, "It was the literal 'Do twice as much to be half as good' thing."

By the time Deb obtained her job with the Forest Service, she had already been through the worst resentment from men in male-dominated jobs, and there were beginning to be more women in the field, so her move from river running to the Forest Service was an easy transition. She could run motors, boats, back trailers and function alone in the wilderness, so it was fairly easy and a lot of fun. She advised, "I have seen a lot of women struggle in male-dominated jobs. I think the key is preparation. As in anything, make sure you have the tools. Also, it's important to know your worth and be able to 'step up' when it is required. It's not always easy, and my feelings

Figure 64. Deb skydiving on her 60th birthday.

have definitely been hurt more than once or twice, but it's the real world and you have to just suck it up and do it. Whenever I would have a bad day, I'd go home and lick my wounds in private, over a cocktail. My philosophy: have a good cry, then get on with it!'

Deb says that she finds it interesting that during the fifteen years she spent working in the Forest Service, she had only two male supervisors, the one that hired her as a seasonal and one in West Virginia. She also finds it interesting that they were the only two supervisors she did not really get along with.

She says that without exception, her female supervisors seemed to grasp that she "got it," and they supported her and let her do her job without feeling the need to try to control her. She said, "I'm not sure if they understood me or just realized I was competent enough to do my job. I don't know the dynamics; it was just interesting. I do know that I have nothing but respect for each of them. Each is a strong and successful woman, and I am grateful to them for being who they are and for what I have learned from them. We developed bonds, and to this day, I am a friend to each one of them."

Deb said that after all the excitement she has had in her life retirement is pretty tame. She said, "I keep thinking that I may return to the workforce in some capacity, but I can't bring myself to work at some boring job. It's a problem, but my options are open."

She is focusing on something she has always wanted to do, learning how to paint watercolors. She said that it is a challenge of a completely different kind, one that will probably take her the rest of her life. She has begun to amass awards for her artwork.

She said, "As to my former exciting life and how I'm coping with retirement? For my 60th birthday, I went skydiving (Figure 64). Two weeks later, I took a helicopter pilot lesson, just to see what it felt like to hold the stick. I'm still here, world."

Barbara Stanley, Forest and Regional Energy Coordinator and Program Manager

A true country girl, Barbara (Figure 65) grew up on a farm about fifteen miles from Connersville, Indiana. There was an elementary through eighth grade school in Bentonville, about 4½ miles away, which she attended with about 120 kids. She went to high school in Connersville. The cities of Indianapolis, Indiana, and Dayton and Cincinnati, Ohio, were each about sixty miles away from the farm.

Both her parents were farmers from way back. The family spent a lot of time outdoors, farming and gardening. Barbara did a lot of farm chores, and grew up driving a tractor. Her family never went camping. She observed that it was the town kids with parents that had indoor jobs who went outdoors for that kind of activity on weekends. Barbara's father was partially disabled, and watched television and read in the evenings after working on the farm all day.

She also spent lots of time outdoors; she packed a lunch and hiked around the farm. She was active in 4-H, especially

Figure 65. Barb Stanley (right), a typical workday on the Craig Ranger District, Alaska.

the conservation projects in weed identification, wildlife and forestry. She won Grand Champion at the county fair for a birdhouse she built and was given a copy of Peterson's Field guide to the Birds. This reinforced her interest in the natural world. She still has the birdhouse — she uses it as a phone stand.

She said she did not take a straight career path. Her undergraduate degree was in piano, organ and harpsichord performance. She had always done music, and was pretty good at it, though she said she had to work harder at it than others. As a child she enjoyed lots of music — she could read notes before words. She began piano lessons at the age of four or five. She got scholarships for music school. During her senior year of college she started thinking how sedentary teaching music would be. She took an elective class in field botany, for which her 4-H experience had given her a good background and skills. Her interest rekindled, she spent a fifth year as an undergraduate to take science classes. Her graduate degree was in conservation and natural resources.

Barbara identifies with the mission of the Forest Service — Caring for the Land and Serving People, and she is passionate about the Tongass National Forest, where she worked for thirty years. What she loved most about her career was carrying out land stewardship, which fit with her farm background. She said, "I love doing what I can to facilitate projects and fill the societal need for renewable, sustainable energy."

Barbara worked for thirty-three and a half years for the Forest Service. She initially developed a mixed bag of valuable experience in a series of temporary positions in different locations between 1979 and 1985. She started as a recreation research technician for the Southeast Forest Experiment Station in Clemson, South Carolina. She assisted project scientists with library research and responded to information requests from Forest Service and university researchers, prepared reports describing research projects and accomplishments and compiled and edited an extensive bibliography dealing with private sector recreation that was later published. She followed that with a similar job with the Department of Recreation and Park Administration at Clemson University.

From mid 1980 through the end of 1983, Barbara worked in technical positions for the Arapaho and Roosevelt National Forests in Fort Collins and Boulder, Colorado.

She compiled recreation use information, assisted the districts with recreation reporting and reviewed proposals for recreation development projects. She collected and analyzed data, conducted literature searches and wrote resource reports and sections of Forest Service management plans.

As a fee collection and compliance officer, Barbara collected, deposited and accounted for fees from campgrounds and was the primary contact with campground hosts. She had lots of contact with diverse forest visitors and school groups, scout troops, church and civic groups, local residents, tourists and families on vacation. Much of her time was spent in enforcing fee system regulations and other forest regulations necessitating frequent cooperation with local law enforcement agencies.

She prepared reports, and supervised volunteer groups. She inspected government quarters for safety hazards, and assisted with hazard tree inventory on developed sites. She gave presentations on low-impact camping techniques and natural history.

Next she headed for Alaska, and never left. She said, "I told my parents when I was a little kid that I was going to run away and go to Alaska. And I did!"

She worked in temporary jobs in Juneau, Alaska on the Tongass National Forest, from mid 1984 through late 1985.

One of her jobs had been project manager for PROJECT TOUCH, a demonstration project that evaluated the use of touch sensitive computer equipment in providing visitor information. Upon her arrival in Alaska, she installed the touch sensitive computer system at a Forest Service information center and designed and entered the visitor information dialog programs. She trained others in operation of the system.

She then did fee collection and compliance work similar to what she had done in Colorado. She also operated the Mendenhall Glacier Visitor Center on fall and winter weekends and gave interpretive presentations to visitors. One of the highlights of her career was living in the visitor center apartment — an incredible place from which to watch the northern lights and listen to wolves. She likes to tell people that she was in charge of the Mendenhall Glacier.

Barbara was responsible for the administration, operation and maintenance of Forest Service developed recreation sites (cabins, picnic areas, campgrounds, trail heads) on the Juneau road system. Duties included inspecting sites to determine the work to be done, obtaining tools and supplies, maintaining signs, bulletin boards and pay stations and arranging for trash removal, toilet pumping, road grading and water testing. She interviewed, selected and trained the recreation maintenance crew and made arrangements for campground hosts.

At the end of 1985, Barbara moved to Ketchikan for her first permanent Forest Service job as a forestry technician in recreation. Just as she had in her earlier temporary positions, she readily took on new tasks, increasing her level of responsibility. As with all Forest Service field jobs in Alaska, travel is mostly by boat or plane (Figure 66).

Figure 66. Barb at work in Twin Otter on way to Hyder, Alaska.

She managed the Ward Lake Recreation Area, including fee campgrounds, several day-use/ picnic areas and a nature trail, and assisted the district resource assistant with other project work, planning and budgeting for the recreation and lands section.

In September 1986, her responsibilities expanded to include lands, minerals and special uses, laying the foundation for the rest of her career. Over the years she became knowledgeable and proficient in a broad spectrum of related disciplines, including recreation and special uses, cultural resources, facilities management, lands management and minerals and energy administration. She served on or provided oversight to ID Teams in all these areas. She took advantage of the opportunity to work in details such as district ranger, forest recreation, lands and minerals staff officer and regional lands specialist. She promoted steadily, moving deftly among district, forest, regional and Washington Office levels in different capacities.

Early on her tasks included processing applications and preparing permits for new special uses, inspecting special uses, mining claim areas and mineral materials sites and reviewing of notices of intent and operating plans for mining.

Later she coordinated and managed a range of projects and programs, such as major hydropower developments, large land exchanges, land selections and conveyances, rural assistance and community development programs, a heritage program that attracted national and international media attention, management of several wilderness areas and an intensive cave inventory and exploration program.

An essential part of Barbara's work included coordination and communication with other agencies, such as the Army Corps of Engineers, Environmental Protection Agency (EPA), State of Alaska, Alaska Division of State Parks, Alaska Department of Natural Resources, the Federal Energy Regulatory Commission (FERC) and tribes.

Barbara was also responsible for understanding and communicating the nuances of the public land laws — several of which are unique to Alaska. She tracked proposed lands legislation that was likely to affect the Tongass National Forest and was the forest contact for questions and information requests. She met with concerned citizens, local environmental groups, Forest Service employees, local government officials and others to answer questions about proposed legislation. She assisted with drafting services for proposed lands legislation, prepared background documents for hearings, reviewed draft testimony and recommended agency positions on legislative proposals.

In 2009, Barbara's broad background culminated in her final promotion before she retired. She became the regional energy program manager, a shared position between the Alaska Regional Office and the Tongass National Forest. This position straddled the line between forest and regional positions. Regional positions are further represented in the *Line Officers* chapter of this book.

As the regional energy coordinator/program manager, Barbara provided leadership and technical expertise for hydropower, wind, geothermal, tidal, oil and gas and energy corridors, with an emphasis on hydropower. Much of her time was spent on the development and interpretation of regional policies and procedures related to hydropower licensing, re-licensing and administration. She worked closely with the Office of General Council, the Washington Office, other regions and externally with utilities, consultants, FERC technical experts and other federal, state, local, tribal and non-government organizations on issues of regional significance, including energy planning. She coordinated the WO review and oversight involvement for the Alaska region projects and program. She coordinated upward reporting, information and report gathering and regional forester responses to FERC. She coordinated, reviewed and/or prepared terms and conditions pursuant to the Federal Power Act and other laws.

At the forest level, her responsibilities were similar to those of the regional program, but the work occurred at the forest. She provided leadership and technical expertise for the forest energy program and advised the forest leadership team on energy-related issues, projects and needed actions. She developed policies and procedures for hydropower project processing and worked closely with district and forest staff to prepare and review responses to project applicants and FERC. She collaborated with the Southeast Conference, Alaska Energy Authority and other organizations in regional energy development planning. She retired in January 2014.

Barbara did not make a conscious decision to turn to a non-traditional career, though she never considered herself to be traditional. She said, "I was independent, with a strong, rebellious streak, would do what I wanted to do. I rebelled against the traditional home and expectations for women, didn't want to be married or have kids."

Amazingly, she could not think of any unusual or specific challenges, but said, "I have worked with older men who thought the agency 'has gone to hell' and admittedly, I've had the same thoughts. I then knew it was time to seriously consider retiring. I do observe young people having opportunities I wish I had had."

Barbara is passionate about the Tongass National Forest and southeast Alaska. She said, "I've lived longer in southeast Alaska than anyplace else; it's home."

Kelly Whitsett, Forest Hydrologist and
Cave and Karst Program Manager

Hydrologist Kelly Whitsett (Figure 67) quipped that she has been studying streams all her life. She was born in St. Louis County and grew up in Rolla, Missouri, a rural community of 10–15,000. The two largest rivers in the country, the Mississippi, and the Missouri bisect the state of Missouri, and the Missouri River merges with the Mississippi in St. Louis. Overall, Missouri has more rivers, lakes and streams than most of the states.

Figure 67. Kelly Whitsett (center) at Forest Service trade show booth.

Kelly's family lived on the water — her parents are both avid fishermen, and when she was a baby her dad loaded her into a backpack and took her fly-fishing with him. She and her brother were always told to play outdoors, whatever the weather. Water activities were Kelly's favorites — her parents could not keep her out of the water, even in winter.

They lived outside the city limits on the edge of hardwood and pine forests. The entire family was active and outdoors oriented. Their home did not have air conditioning, and they took advantage of the abundant waterways to cool off in summer. They spent the hot time of year floating canoes or kayaks down rivers, including an annual three days long Father's Day float trip, and they water skied frequently at the Lake of the Ozarks.

Fishing was her parents' favorite water activity. They chose quiet upstream locations away from the more populous downstream areas for floating, fishing and camping. Camping and hiking were their main activities during early spring and late fall.

Kelly was also a competitive swimmer and lifeguard, and taught swimming in high school. Recently, she and her brother purchased a boat together, and Kelly has turned to wake boarding, which she said is a little easier on the body than water skiing.

Kelly comes from a family of scientists and teachers. Her father taught high school physics and calculus, so it not surprising that her favorite school subject was math. She was in math club, and enjoyed chemistry and physics classes.

Kelly was ultimately attracted to her profession because of her love of and interest in water, though she took a bit of a circuitous route to getting there. She was thirteen and already exploring her options, when her dad happened to enroll in a college geology class and she accompanied him on a field trip.

Besides water, Kelly had a second love of collecting rocks. Until the field trip with her dad she had not known that geology was a subject. There was lots of rock talk during the trip, and for the first time Kelly said, "I realized that geology was something you could do for a living!" She also entertained an interest in being an astronaut and her dad told her that NASA hires geologists.

As a high school senior Kelly decided to study geology at the School of Mines in Rolla, the oldest of its kind in the country. Her emphasis was in geology, which included study of karsts, limestone landscapes that have sinkholes, sinking streams, springs, underground rivers and caves. It was fascinating. She was a member of a spelunkers club, which explored caves and did such work as mapping underground streams.

Kelly received her Bachelor of Science degree from the school of mines in December 1999. Most of the available jobs then would have been in oil and gas, which she had determined was not for her. She only saw hydrology as the interesting component of her mines education, so she entered a masters program in hydrogeology in Fayetteville, Arkansas, and received that degree in May 2002.

Two of Kelly's professors worked for the U.S. Geological Survey (USGS), and they advised her that it was hard to get a job with that agency. They encouraged her to apply for all federal jobs in hydrology in any places she was willing to live. She applied to all agencies and lived at home while waiting for a job offer. She accepted a part-time day care job at a children's hospital and had attended her first day of orientation training. The next day the Forest Service Region 8 Office in Atlanta, Georgia offered her a hydrology trainee position in their training program. The job was in Hot Springs, Arkansas. It was a two-year appointment, which required Kelly to sign a mobility agreement, saying she would transfer to an available permanent position anywhere after two years. The purpose of the program was to train people in positions mainly in hydrology, soils and botany that were being lost to retirements. Recruits were fast-tracked to be placed in the positions vacated by retirees. The end of Kelly's two years occurred during the Clinton-Bush transition, when due to uncertainty about the economy, Forest Service employees deferred their retirements, eliminating the anticipated placement opportunities for the trainees.

Kelly was left with two choices; apply for all GS-9/11 hydrology positions in the Forest Service or leave the Forest Service. She applied everywhere, whether she thought she would like it or not. She accepted an early offer from the Plumas National Forest as district hydrologist on the Feather River Ranger District in Oroville, California. It was the farthest she had ever been from home.

On the Feather River district Kelly supervised a few employees who helped her with responsibilities for the management of hydrology and soils. The main part of the job was support for the demanding HFQLG program, collecting data and preparing environmental analyses for fuels and timber projects. She also did landscape design for stream morphology, working with engineers and aquatics biologists. Watershed restoration was a smaller part of her duties at first, but she loved that part and grew it, partly by applying successfully for

Figure 68. Kelly at home in the water with companions Chelsea and Annie (background).

matching grant money. Her budget for that work was about $200,000 per year at first, which she grew to $1,000,000 in her final year on the district. The dollars paid for meadow restorations, natural channel design that fixed eroding stream banks and replacing stream crossings where there were barriers to aquatic organism passage. There was lots of roads work to ensure projects requiring roads would meet Best Management Practices (BMPs) for watershed health, by improving streams and controlling how much sediment would enter streams as a result of proposed projects.

After 5½ years of hard work, Kelly was able to move back to her home in Rolla, Missouri with a promotion to her dream job as the Mark Twain National Forest Hydrologist and Cave and Karst Program Manager. She now lives with her two dogs (water dogs, of course) that she takes along for water-related recreation (figures 68 and 69).

Figure 69. Kelly and Annie catch a small mouth bass from the canoe.

She bought a house on three acres outside the city limits and keeps a large garden, preserves her own food and even makes her own dog food. She is happily independent, not averse to marriage, but she said, "I never met my soul mate, and I do not want to settle." She is close to her family and sees them often. Her parents retired at the Lake of the Ozarks that is about 1½ hours away, and her brother and his family is in St. Louis, about 1¾ hours in the other direction. Kelly has two nieces and a new nephew. Her nieces like being in water, collecting bugs and rocks.

Kelly's job on the Mark Twain encompasses a complex of duties, managing surface and ground water, karst and caves and BMPs throughout the forest.

The program manager position had been vacant since 1988, so Kelly had her work cut out for her. The forest has a large prescribed burning program that has experienced a lot of public opposition. Part of Kelly's job is to support timber and fire, conducting NEPA analyses of effects to hydrological resources from logging and burning projects.

Her highest priority is planning and implementation of projects to improve the watershed landscape (Figure 70).

Figure 70. Kelly (center of map) planning a restoration project.

She oversees management of hydrology for three priority watersheds, covering 1.6 million acres. Her jurisdiction includes miles of streams and scenic rivers and seven wilderness areas. She travels a lot around the forest in order to consult with the districts.

The Mark Twain National Forest has 760 caves, the most of any national forest, and Kelly oversees their management. That work involves surveying for threatened and endangered species, and in many cases gating caves to protect those species.

Kelly is on a regional technical systems team that has developed a program to return large woody debris (downed trees, e.g.) to streams that are missing trees that would naturally have been there if not for land clearing that has occurred over time. The woody matter, or lack of it, influences stream morphology and habitat for aquatic species.

Being an extrovert, Kelly enjoys building watershed restoration partnerships. The largest to date is with The Nature Conservancy, Missouri Chapter, who wants to share a watershed restoration designer position with the Forest Service.

Aquatic biodiversity on the Mark Twain is the highest in the region. There are a lot of threatened and endangered species, which leads to a lot of natural channel design work. There are related issues with ditch systems. Bridges act as dams, and bridge replacement projects improve passage for aquatic species.

Another important partnership Kelly has formed is with the Ozark Chapter of the Cave Research Foundation (CRS). Volunteers work throughout the 760 caves on the forest, mapping caves and conducting surveys for biological and geological resources. They monitor the caves for vandalism and remove trash. Many caves are also Native American tribal sacred sites containing cultural resources.

Administrative duties, NEPA, the inevitable reports and applying for grants keep Kelly office-bound for a fair percentage of her time. She makes a point of getting to the field one to three days per week, depending on the weather (Figure 71). She enjoys a yearlong field season, generally spending more river time in her canoe in summer and woods time in winter.

Figure 71. Kelly at large woody structure project to protect stream bank and create fish habitat.

Kelly never cared about differences between female and male roles. She said, "I didn't grow up that way, gender roles were not a big deal in my family. If I thought I could do something boys liked, I did it. But I also liked being a girl and wearing dresses. I had a younger brother who I drug outside when he was scared to go, to do things like digging worms for fishing. I led him into a love of the outdoors. When it came to my career choice, I only knew I wanted a field job. I did notice that there was a mix of male and female students in school, but few female professors, and sometimes at work I was the only female in the woods. I have noticed more women in the natural resources field recently. Three of four students who work for me are women." She said she thinks women help bring a balance to the work place, that they have different viewpoints or ways of looking at things and often are better at multi-tasking. She said, "The work place needs the diversity of men and women and the different ways we all approach the work. It is very important in the Forest Service to work together as a team to accomplish the program of work. I currently work in an office that works great together as a team, and there is a nice mix of about 50/50 men and women in the program manager roles."

She has never faced sexual discrimination against her. Rather, being "blessed" with a youthful appearance, she had a hard time getting people to take her seriously as a competent supervisor and natural resource professional early in her career. People had a hard time being questioned by her. Personnel issues were a lot to handle. She had an employee who would not take direction from her and another who filed a grievance against her. She said, "I worked very hard to get to where I was, and I felt disrespected. It was so bad for awhile that I would literally cry every morning on my way to work." She actually considered quitting and going back to school for a doctoral degree. She went so far as to apply for a PhD, and then accepted the status quo and concentrated on the good experience she could get, deciding that work experience was her better option.

Figure 72. Kelly wake boarding — it's all about the water.

Forest-Level Natural Resource Professionals

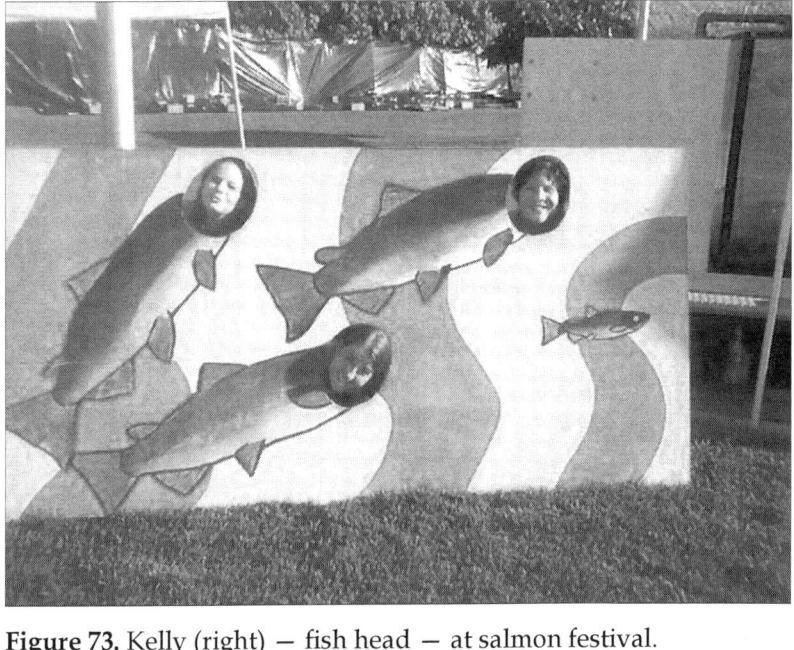

Figure 73. Kelly (right) — fish head — at salmon festival.

Another type of challenge was having a line officer press her to change the data she had collected so that the line officer could make the desired decision to go forward with a proposed project as planned. Kelly stuck to her guns under pressure.

As of this writing, Kelly was about fourteen years into her career. She still considers the geology field trip she attended with her father when she was thirteen to be one of her most extraordinary life experiences, "the thing that started shaping it all." Learning about watershed restoration while working in California was another "eye-opening" experience. Watershed restoration was not a common practice in Missouri, and California was a leader in the process, so she got to learn from the best.

Growing up being in woods and on rivers, creeks and streams, Kelly is happiest there (figures 72 and 73). She feels that water is our most important resource and that it does not always get the importance it deserves. She said, "My job is a lifestyle." She enjoys working with students and has developed a program agreement with the university in which she trains and employees about four students at a time. She said her passion is "to improve upon what people before did to make the land better, to continue the evolution so it is there for the next generation, and to give back the mentoring I have received."

Chapter 5

Recreation Management Specialists

National forests and grasslands are the largest providers of outdoor recreation opportunities in the United States.[33] People from all over the world visit them for hiking, camping, swimming, boating, fishing, skiing and more.

Of all Forest Service employees, recreation managers probably engage in the most interaction with the public, and so have a special role in connecting people to the land. They balance the use of forests for recreation with the need to protect ecosystems. The duties of the employees who work in recreation are as varied as the people who play in our national forests.

Employees at all levels manage the recreation resource. Technicians maintain trails, campgrounds and picnic areas, and they patrol developed and undeveloped recreation areas to enforce rules and educate the public about using the land responsibly. This latter responsibility may include the foot or horseback patrolling of wilderness areas, designated parts of forests that are intended for use that is lighter on the land in order to preserve its more pristine qualities.

Recreation program managers at the district level and above look at the bigger picture, manage budgets, plan recreation maintenance and development, create interpretive sites and programs and establish and maintain partnerships, including those with volunteers who help maintain trails and facilities.

The managers rely on research by Forest Service scientists, personal experience and input from recreation employees and the public to guide planning and decisions about what kinds of recreation to provide, based on public demand and the nature of the geographic area being considered, whether that is an urban or wild setting, for example. They provide for recreational needs of people from all walks of life, considering age, ethnicity, physical ability and geographic location. They interact with people from diverse cultures whose needs differ from each other. They determine how much use a land area can support without damage, and how much money a particular use may be worth. On ID teams, they consider the effects of other types of projects, such as timber sales, on the recreation resource, while seeking to balance recreation needs with other resource needs.

Compromise is sometimes required to find that balance among all the affected resources being considered. For instance, the recreation management specialist may need to remind timber sale administrators that a buffer needs to be maintained along hiking trails. While Forest Service management activities are evident across national forest lands, they need to be balanced with recreation use. Hikers will prefer the experience of hiking through a forest rather than through timber management activities. Sometimes the recreation resource is sacrificed, an example of that being a narrow corridor between a lake and a highway where bicycle trail visuals are already breached by overhead power lines. A proposed addition of a buried pipeline corridor will further diminish the values that bicyclists enjoy. Given no other alternatives, a decision to issue the pipeline permit along the route would further diminish the visuals of the bicycle route.

Recreation management specialists have a role in repairing, renovating, replacing, expanding or building new facilities to ensure they meet Forest Service standards for fitting aesthetically into their rustic landscapes and that they are built and operated sustainably. They maintain the character of important historical sites. They rely on the services of architects, landscape architects, archaeologists and engineers.

Interwoven into all recreational uses is protection of historical, or cultural resources such as Native American artifacts and sacred sites. To that end, recreation managers foster relationships with local tribes and work closely with tribal governments.

Many recreation management specialists have a broad spectrum of duties to juggle. In addition to strict recreation management, they may work for or alongside lands management officers, and have additional responsibilities related to land exchanges and sales, special use permits for uses such as campgrounds and ski resorts and oversight and coordination of hydroelectric projects that provide water-based recreation opportunities on national forest lands.

Clearly there is a place in the Forest Service for those employees interested in some aspect of recreation management. Many of the women represented in this book spent a part of their careers in recreation management. For the three featured in this chapter, recreation was the main focus of their career.

Sonja Hoie, Recreation Program Manager

Sonja (Figure 74) was born in Anchorage, Alaska, and then moved to Yosemite National Park when she was four. Her mother worked for the military as a civilian in Alaska, then spent the rest of her career as a National Park Service (NPS) employee in Yosemite.

Figure 74. Sonja Hoie.

In such magnificent surroundings, Sonja was naturally outside as much as she could be. As a younger child she recalls summers playing with friends until dark.

Her elementary school had a ski program in the winter starting in the early grades. They would go to school early on Wednesday, and then ride a bus to the ski area in the afternoon for ski lessons. If they didn't ski they went to the ice rink and played crack the whip and trains, going home with bruised knees and elbows.

As an older child Sonja and a gang of friends "pushed our parental boundaries and secretly explored caves, swam in forbidden swimming holes and made forts in the forest."

Play in the park was inclusive of boys and girls. They explored caves, had acorn or snowball fights, rode bikes and had bike obstacle courses, played crack the whip at the ice rink and skied together. Sonja said, "Given growing up in a national park, perhaps we operated outside what would be considered typical girls' activities. My girlfriend and I built a secret fort on a cliff side. We went exploring and hiking. I did, however make note and was resentful that the Boy Scouts had adventurous activities (overnight hikes) while the Girl Scouts made slippers for the hospital or carried a bag lunch on a day hike."

High school years were spent hiking, backpacking and camping. Sonja and her mom hiked, and her aunt and uncle and cousins would come to the park to hike and take extended pack trips. They skied, camped and enjoyed the outdoors at every opportunity.

Her favorite school subjects were art, English and biology. She recalled being interested in science in the early grades, particularly the natural environment that surrounded her. She said, "Math, not so much, but I think much of that was linked to having to recite multiplication tables in front of the class."

In the summer Sonja's mother took her to the family farm in Iowa for a month where she would feed the calf, pester her uncle to let her milk the cow and help him with chores, pick apples, help in the garden and find ways to entertain herself on the farm.

Sonja did not get to know her father well, because he died when she was young. From what she does know of him, she said, "I believe his adventurous spirit in coming to the United States after World War II may have influenced my career choice." He was in the Norwegian Resistance movement. He immigrated to the United States after the war, worked in Alaska installing remote radar stations for the military and eventually opened a saw shop. She further commented, "My mother was also adventurous and independent. She and my aunt left their family's Iowa farm, went to business school and worked in Baltimore, Seattle and Yosemite. For these two young women to leave the farm and pursue careers was truly adventurous in that era. Mom worked in Yosemite, then Alaska, where she met my father, then returned to Yosemite with me."

Sonja wanted to work in public land management and have a part in care-taking them for the public good. She wanted to be outside. She said that the fact that her career was non-traditional was not the attraction; it just was not an issue. She said, "Not that there weren't plenty of challenges due to the male-dominated culture. That it was important work and was outdoors was the draw."

Her first "non-traditional" job was with the Park Service during the early 1970s, working in a kiosk at the entrance station. Males had traditionally held even that job. She reflected, "I guess it did have an element that would have been viewed back then as non-traditional for a woman — I once had to perform CPR on an accident victim near the entrance station, who as it turned out, had already died."

For women of her generation, wanting the challenge, including the physical challenges and being told you could not do something because you are female could be difficult to overcome. It was a challenge to Sonja having "fatherly" supervisors who thought she needed their protection, which she resented. She thought the paternal supervisory style limited opportunities for women.

She had a pregnant co-worker who was actually told by her supervisor to hide when the park superintendent visited, so that he would not notice that she was pregnant.

Sonja said she experienced more male resentment when she worked for the Park Service, than in her Forest Service career. She speculated that the reason could have been due to it being during earlier times, the particular individuals, or the workplace culture.

She experienced sexual harassment and bullying from supervisors in two different jobs when she worked for the Forest Service. In both situations she chose to follow informal channels, bringing the problem to the attention of the district ranger and the forest supervisor. The district ranger immediately addressed the sexual harassment problem and found an acceptable solution. The forest supervisor, in his inexperience, actually exacerbated the situation. The immediate supervisor's eventual retirement solved the issue. During her Forest Service career, she saw other women that chose to transfer due to resentment and lack of support from supervisors to remedy a situation that was clearly hostile and not addressed due to the proverbial "good 'ole boys" network. She said, "These women eventually made these moves work for them — a compliment to their resilience." Overall Sonja describes her forest service experience as positive. She stayed with the agency because the positives overrode the negatives. She said, "Even when things were bad, I had a higher purpose." She said that she never felt deliberately excluded or asked along as a token, and that the times she felt she did not fit in during her Forest Service career were primarily linked to being a creative thinker in an organization that was very traditional. She said, "That environment changed over the years to become a workplace that valued more innovative thinking."

She was a single mom, and had incredible agency support. Single parenting was not really a challenge for her, though she observed that a challenge for women of her era integrating into a male workplace included having to demonstrate, many times, that you could do the job and balance family needs. In Sonja's case flextime was very helpful and coworkers were very supportive. The community of Pagosa Springs, Colorado, where she lived while her daughter was young, was also supportive. The school district had a program that kept the same kids together for three years, and the parents did eight hours of volunteer work at the school each month, which brought them all together.

Having a western mindset, cultural differences when she transferred to the Ozarks in Missouri presented challenges. With no family, no spouse, no local church, she did not fit. A lot of the public there did not know who the Forest Service was and what they do. Once she was mistaken as a United Parcel Service employee when she was in uniform.

The cultural differences were not as pronounced when she worked in the Midwest, but she found it odd that the perception there was that western forests are superior to mid-western.

Over a thirty year career with both the Park Service and Forest Service, Sonja enjoyed the variety of her work in Yosemite National Park, Grand Canyon National Park and the San Juan, Mark Twain and Chippewa National Forests. Her day-to-day was everything from marking a new trail route, working in campground and entrance stations, leading a pack mule across a precarious route in the Grand Canyon or riding horses to clear wilderness trails with a cross-cut saw. She issued special use permits, wrote violation notices and appeared in court. She made the case for a new position, compiled stimulus project proposals, squeezed every penny out of her recreation budget to get the job done, partnered with the public on scenic byways, worked on interpretive plans, mentored younger employees and more. Some of her responsibilities are detailed in descriptions of her specific positions below.

She got a bachelors degree in environmental science from Chico State University in California in 1977. She first studied earth science and environmental studies at the University of California (UC) at Santa Cruz for three years between 1970 and 1974, but found she wanted more structure than UC Santa Cruz had. She also did field study courses in geology and botany at UC Davis, in between quarters at UC Santa Cruz.

Her first job related to her studies was as a park ranger for the NPS in Yosemite. She worked first seasonally and then

Figure 75. Sonja, early career — NPS's first full-time female backcountry ranger — Grand Canyon.

permanently between 1972 and 1978. She was an entrance station supervisor, and she did campground fee collections and provided visitor information. She participated in the bear management program.

Next, from 1978 through 1981, she was an Inner Canyon ranger for the Park Service at the Grand Canyon National Park in Arizona. She was the first full-time female backcountry ranger there (Figure 75).

Within the park, she worked at Indian Gardens, Phantom Ranch and Tuweep. She was stationed in areas accessible only by mule or hiking. It was crucial to be aware at all times of environmental dangers, such as the very real possibility of falling off a cliff, or being bitten by a rattle snake. She had to manage the mule she was riding and a pack mule. She had to keep herself and others safe. The job included intensive public contact, law enforcement, campground and trail maintenance, resource restoration and she was an EMT. Other duties included developing an operations and maintenance manual and researching and compiling a herbarium. She serviced diesel generators and monitored pumps providing water to the South Rim of the canyon.

In 1985, Sonja went to work for the Forest Service on the San Juan National Forest in Dolores, Colorado. She was the recreation staff officer for the Dolores district.

She developed capital investment project narratives and participated in team planning for recreation sites. She managed the district recreation program encompassing campgrounds, wilderness, law enforcement and trails, including planning snowmobile and hiking trails. She developed a guard station rental program for public use.

As a Southwest Council member, she worked to establish networks with Jicarilla, Southern Ute and Ute Mountain Indian tribes.

She was the district liaison for San Juan Mountains Association, a San Juan National Forest friends group.

She was a team member in planning management of two new reservoir based recreation areas with mixed jurisdictions. She determined fee structure, boat patrol procedures, lakeshore management, participated in co-op search and rescue and in developing law enforcement plans.

Sonja moved to Pagosa Springs, continued working in recreation and then switched jobs to the lands and special uses administrator. In that job she processed and administered permits for various proposed uses of national forest land by non-forest entities, using a data system to generate permits and billing. Permitted uses included communication sites, road easements, recreation cabins, utilities, snowplowing, pastures, hay cultivation, treasure hunting, rights-of-way, spring developments and recreation events. She prepared environmental effects documentation for related project work, using knowledge of various natural resource disciplines. She performed preliminary negotiations in land exchange proposals or donations, and investigated and resolved encroachments (trespass).

Figure 76. Sonja monitoring visitor use on Chippewa National Forest.

While working both on the Dolores and Pagosa Districts she performed collateral duties as an Equal Employment Opportunity Counselor.

In 2004, she was promoted to the supervisory natural resource specialist position on the Willow Springs district of the Mark Twain National Forest in Ava, Missouri. She exclaimed, "What a change from Pagosa!" There she administered a motorcycle and ATV area, three wildernesses, developed campgrounds and picnic areas, motorized and non-motorized trails systems, a scenic byway and outfitter guides.

She was on the Wonders of the Outdoor World planning committee and instructor for an annual environmental education event sponsored by Bass Pro, Missouri Department of Conservation, Wonders of Wildlife Museum and federal partners.

She developed a partnership with the Chamber of Commerce to maintain a picnic area for their annual event and a developed a brochure to promote the scenic byway. She worked with an equestrian group to develop riding opportunities and volunteer efforts to maintain trails.

2006 brought another promotion, to the forest recreation program manager for the Chippewa National Forest in Bemidji, Minnesota. Sonja was the manager of the forest recreation program; twenty-one campgrounds, 700 miles of motorized and non-motorized trails, water accesses, dispersed recreation opportunities, four scenic byways, historic properties, partnerships and agreements (Figure 76).

She provided direction and advice to district recreation program managers while building a team environment and organizational efficiencies. She engaged the forest in assessing needs and developing priorities for a 1.3-million-dollar annual budget. She reduced costs and improved operating efficiencies by initiating new contracts, introducing use of Student Conservation Association naturalists and use of the Leech Lake Band of Ojibwe temporary employment program.

She developed communication links between external partners, forest leadership and the eastern regional office (Figure 77).

Figure 77. Sonja — on a field day with trails volunteers takes a moment to enjoy the milkweed.

Figure 78. Sonja mends fence on her Iowa farm.

She provided interpretive planning and design as a team member for Edge of the Wilderness Discovery Center. She reviewed and edited scenic byway interpretive master plans. She proposed, planned and implemented Recovery Act projects in trail maintenance and recreation facility replacements.

Sonja retired in 2010. With the luxury of being retired she is turning to the things she never had enough time for, weaving, spinning, botanical drawing. She said, "Those are my passions, along with traveling to other wild places in the world."

She moved to the family farm in Iowa, the farm that has been in her family for over 150 years. She built a new home on the property. She said, "I will take much of what I've been fortunate to learn from working for the Park Service and Forest

Service to manage the farm in a sustainable manner, and to do needed repairs (Figure 78)."

She said that her career gave her confidence in many areas of her life. She said, "In the Forest Service I learned to build fence, care for horses and run a chain saw. I had training in landscape management, and got to look over plans for campgrounds and visitor centers. As part of interdisciplinary planning teams, I gained knowledge of other disciplines including range and wildlife management and preventing or treating stream erosion. All of these help me in the complex endeavor of managing a farm."

She said her most memorable work experience was her time in the Grand Canyon, given the place and the job. It was always extraordinary, too, working in beautiful surroundings in the woods and mountains for the Forest Service, doing things like cutting trees off trails. She said, "I loved being a steward of public lands, discovering and accomplishing what was important to the American public."

Sonja finished by saying, "I made the right career choice — I did important, satisfying and challenging work. The Park Service and Forest Service provided opportunities for career and personal growth. I have a skill set that is serving me well in this next chapter in my life. There are many things that concern me regarding the future of public lands, but now I am focused on the things ahead of me in my next life."

∽

Joan Kluwe, Recreation Planner

Joan (Figure 79) is from Alton, Illinois, a town of 40,000 on the Mississippi river across from St. Louis that is surrounded by limestone bluffs, mixed hardwood forests and farms.

As a child she liked outdoor activities, including Girl Scouts, camping and canoe trips. Her connection with the natural environment developed through Girl Scouts. She liked science and reading, but less reading as she got older, as it was more passive. She liked most school subjects and did well in them. Science was one of her favorite subjects, and she recalls

Figure 79. Joan Kluwe at Point Barrow — northernmost point in Alaska.

enjoying geometry and statistics and graphs in elementary math classes. In high school she liked biology, chemistry and physiology. Now she enjoys more physical, action-oriented activities. Even her static activities are more action-oriented, such as quilting, which involves producing a product versus passively reading.

Joan has an older brother and a younger brother and sister. Her family was not particularly outdoor-oriented, but they had a large garden and woodlot on five acres when she was young. They grew their own Christmas trees there and produced firewood.

Family activities did include fishing in nearby ponds and a summer vacation to a state park lodge to go trout fishing. There was a gender bias in those days — her brothers hunted with their father, but girls did not hunt. Girls could fish, and Joan joined in the fishing. She often felt more affinity with boys than girls — girls' activities didn't bore her but she said, "Some of the boys' activities were just more exciting!"

Girl Scout activities attracted Joan to her profession. Her father's activities also influenced her. He was a skilled taxidermist, which he practiced as a sideline. Her deep fascination with animals arose from exposure to her father's taxidermy hobby.

Joan did not deliberately turn to a non-traditional career. Male-dominance was more noticeable in college, where men outnumbered women in the classes. She majored in forestry with an emphasis in outdoor recreation management. Her Forest Service job following college had more gender diversity than in school. There also were more women than men in recreation jobs. As she continued into recreation management and then got into planning and environmental compliance, there was good representation of women in the recreation field, and sometimes women dominated the field.

By the time Joan entered the Forest Service as a young professional, there was not a strong gender bias, but older women

Figure 80. Joan hiking Crow Pass.

that she worked with told her stories of having been told by Forest Service recruiters in the early 1960s not to bother applying for Forest Service jobs, unless they wanted to be a clerk. Joan said, "The Forest Service has changed over the years, from overtly excluding women in the early decades of the agency to women in leadership roles today. I see fewer gender challenges in today's agency."

The physical aspect of site reconnaissance and trail work, hauling heavy packs up and down steep hills, was challenging but enjoyable (Figure 80). The static element in planning allowed Joan to transition out of some of the heavily physical work. She wants to be able to have time and energy for personal recreation and manage the physical toll on her body.

Another challenge of working for the Forest Service in a small isolated community was the lack of opportunities for varied social interaction. Joan lived and worked with the same small group of people. She turned to sea kayaking as a means to get away and be alone to renew (figures 81 and 82). She also enjoyed the company of her dog, Cody (Figure 83).

She has worked for agencies other than the Forest Service, and said, "I liked the Forest Service best, getting paid to go where others pay to go. I saw some amazing places during my Forest Service career (figures 84 and 85). My experiences with the agency were foundational for my career — my technical skills were expanded as well as my self-confidence."

She worked in Colorado as a wilderness ranger, supervising a crew who patrolled campsites, cleaned up and educated the public about Leave No Trace camping (figures 86 and 87).

Next she worked as a wild and scenic river planner, then recreation planner on the Mount Hood/Clackamas, Zigzag and Estacada Ranger Districts in Oregon as a recreation project construction manager.

She was planning to start a Ph.D. program in Idaho and was accepted to the university, but then she saw an ad for a job in Alaska, posted by an acquaintance there, Barb Stanley. Joan called Barb to ask about the position, and Barb convinced her to apply. She was offered the job, but first had to fly to Alaska

Figures 81 and 82. Joan in her sea kayak in Prince William Sound.

Figure 83. Joan's favorite photo with her best pal Cody — sunning on her deck.

Figure 84. Joan at Stoney Ridge — view of Denali.

Figure 85. Joan at play — Alaska Range, Kesugi Ridge.

to check it out. She had once had a challenging assignment in south Texas, and promised herself no more blind moves. She decided to move to Alaska for two years and to save money for graduate school.

She went to southeast Alaska in 1994, as a recreation planner. Her duties included field visits to look at proposed timber harvest areas and estimating impacts to recreation sites and activities. Her work also included wilderness management, recreation facility management and construction management for the district's first developed campground. She prepared environmental assessments and environmental impact statements relative to recreation issues. She eventually did start graduate school in the fall of 1997, but her home remained in Alaska.

One of Joan's most extraordinary experiences occurred in southeast Alaska, where she had developed a voice deer call. She had just pushed off her boat from the shore but not yet started the motor. From ten to twenty yards off the beach she did her deer call and heard loud thrashing/crashing as a mamma deer came rushing out of the brush. Another time when she called, three bears came on the beach. Calling the deer also alerted bears that were out looking for injured deer.

Figure 86. Joan making sand angels — geared up for Colorado River trip — leisure time.

Figure 87. Joan pointing out Ooh Aah Point, Grand Canyon.

When living on Prince of Wales Island, Joan often went cross-country skiing. One time she went alone on a road with fresh, trackless snow. She skied out and back in the same tracks. On the way back, she could hear wolves off in the distance and then she noticed that there were tracks from a wolf that had curiously followed her for a while as she had skied out. The tracks showed the wolf turned around just ahead of her return and then veered off into the woods. She only skied alone along the road. When going to steeper or more remote areas, she went with a partner for safety.

Joan left the Forest Service in 1999, to accept a Fulbright Scholarship and continue work on her Ph.D. The Forest Service was downsizing at the time and recreation management was being severely cut. Several of the forest/regional managers had no background in recreation and their leadership disheartened her. In addition, she was looking for more of an intellectual challenge than she was routinely finding in her work. She at first offered to resign to accommodate the time to pursue her research, but her supervisor assured her that they could work out a leave of absence. Ultimately, however, the forest supervisor denied the leave of absence, and Joan did resign. While she experienced a "roller coaster of emotions" over her decision, the choice between accepting the prestigious Fulbright Scholarship and the small-mindedness of forest management was easy. She said, "I just felt it was time to explore something new, before becoming truly dissatisfied with the Forest Service and becoming a crabby person."

She completed her Ph.D. in social science and natural resources, with her dissertation examining land use conflicts between subsistence and recreation uses in Alaska and Finland. She lived in Roveneini, Finland for nine months to conduct part of her research. After completing her Ph.D. she went to work for URS Corporation, a global engineering and environmental consulting firm with over 100 employees in Alaska. She was a senior planner and worked in preparing environmental impact statements, environmental assessments and community and regional plans for federal, state and private sector clients all across the state.

After a dozen years in private consulting, she returned to federal service as the Environmental Protection Specialist with the Alaska Region of the National Park Service in early 2016. She said, "I feel like this is where I belong; I have returned to my roots," She works with environmental planning and compliance for parks in Alaska. Joan has enjoyed sharing her knowledge and experience with park staff; she is mentoring several colleagues working on National Environmental Policy Act compliance.

Joan is grateful for the diversity of her career opportunities in federal service, consulting and academia. Each has helped to broaden her understanding and appreciation of natural resource management, and she feels she has been able to make positive contributions to the agencies, clients and research.

Cornelia (Connie) Lane, Recreation Program Manager

Born and raised in Mobile, Alabama, Connie (Figure 88) grew up in St. Mary's Girls Home, a catholic orphanage for girls. Her mother abandoned the family when Connie was young. Her dad was deaf, and the state took the children away. They became wards of the Catholic Church. Her brother was taken to a farm to work most of his life and it turned out the family that raised him was really good to him and treated him like a son.

The girls at the orphanage spent most of their time in the outdoors. The Sisters, the Daughters of Charity, believed that the outdoors was healthy for both body and soul. Their spring through fall quarters were at Battles Wharf, Alabama where they spent sunup to sundown swimming, water skiing, seining for seafood for their supper every day, canoeing, hiking, archery and basking in the sun on Mobile Bay. It was a life filled with adventure, creativity and imagination.

The girls lived in a mansion on Mobile Bay. A doctor donated the mansion to the Catholic Church, who found its best use to be as an orphanage for girls, run by the sisters.

Figure 88. Connie Lane — Mesa Ranger District.

The girls never really camped out. They slept outdoors, basically, as their bedroom was in a u-shaped screened in porch, with just a roof and screen to keep the elements out and to ease the heat of summer months with flowing summer breezes. They could see the stars at night, as the girls' home was in a very rural area with no streetlights. Every night was a campout.

They water-skied almost every day. Connie started water skiing when she was five years old. They had the old wooden skis back then and it was not long before she graduated to one ski. She said, "I think the girls at the home were some of the best skiers in the state. They certainly would catch the eye of other boaters in the area."

Connie reflected on the core value of religion that was embedded into every aspect of the girls' day:

> We were taught to help one another no matter what. We were a team brought together by God. We were taught that no matter our differences, inside we were all the same. We

grew to value each other's talents and strengths. We learned to be individuals uniting when needed. We learned to treat others as we wanted to be treated. We learned that when we were set free that life would not be easy, however if we were on the right path we'd know it and our faith would lead the way. If we strayed we would know that too, because we would feel the lessons we were meant to learn. Through the years I have proved that fact to be correct.

In the 1970's, when Connie graduated from high school, the girls were "set free — you can imagine how scared a feeling that was." Connie had a boyfriend that had gone to the Grand Canyon to work. She saved up enough money to visit him. She got to go down the Colorado River with him and the deal was that she had to work for the trip. So she swamped, which meant she was at the mercy of the river guides. She was basically their "gopher." She considers it the best time of her life that "for once, set my spirit free."

She did her first hike out of the Bright Angel Trail, another experience she will never forget. She said, "We didn't have real rocks at Battles Wharf. I don't even think I'd ever seen a real rock, much less canyons and white water." Her adventurous spirit wanted more of the western outdoors. She returned home after that summer determined to return to the west, which she did within a year. Her boyfriend moved to Idaho where his family started their own outfitter-guide business. She worked for them and for other outfitters in the area. She had found her new home and being outdoors was "my only love in life." She wanted a part in the future of the great western outdoors, and so the journey began.

After graduating from Bishop Toolen High School in Mobile, Alabama in 1970, Connie attended Mississippi College for Women for a year. She worked for a while for real estate and title companies, then in 1971, moved to Idaho where she worked for several outfitters and eventually went into partnership in the business. After leaving that work ten years later, she worked as a secretary in Boise, Idaho until 1986, when she moved to Arizona where she eventually found her way

Figure 89. Connie (right) and colleague David Slan — Tonto National Forest.

to work with the Forest Service on the Tonto National Forest (Figure 89).

What she loved most about that work is that it left her feeling that "somehow, somewhere, I have made a difference." She loved being involved in "generation" projects that will leave untrammeled experiences for our youth to follow. She finds it rewarding to know that she has a part in educating and bringing up new leaders with strong ethics and stewardship for our public lands, in providing places for flora and fauna that we experienced in our lifetime to survive. She marveled, "I really enjoy the multiple use concept, and being a leader in teaching responsible natural resource management is key to my happiness now; to not only leave behind a personal legacy, but a generational legacy that has done well. The diversity of our jobs is incredible (figures 90, 91 and 92).

Being a part of the Forest Service family is indescribable for a woman who has never had a real family. The Forest Service is my family."

She started in clerical work, and while raising her daughter on her own, started attending college classes, one class at

Figure 90. Connie packing out of the Superstition Mountains.

Figure 91. Connie boating home from cleaning Canyon Lake recreation sites.

Recreation Management Specialists

Figure 92. Connie at the Lost Dutchman Days Parade, Tonto National Forest.

a time. In 1996, she graduated from Arizona State University, Summa Cum Laude — highest honors. Her career advanced from there into assisting with management of the Mesa district's recreation, lands and minerals program, to a staff officer position on the neighboring Globe district, a detail as a district ranger and finally a forest program manager position in recreation, lands and special uses on the Coronado National Forest (Figure 93).

She retired in 2015. She had worked fire assignments during summer seasons, which she still does during fire season. She is currently a planning section chief for a Type 1 Southwest area team.

Connie never thought about her job being non-traditional until she became a staff member. She realized then that she was the only female staff officer at meetings. Then it became apparent to her that more females are needed in the management world.

It was a challenge for Connie trying to move up in the same region where she began her career, the same roadblock many

Figure 93. Connie (front, 2nd from left) and Forest Leadership Team, Coronado National Forest.

agency employees face. Some, due to family matters and personal reasons, are not able to move around the nation. If they are not willing to do that then it is difficult to advance their career. Connie said, "To me that is unfair, and a lot of knowledge gets lost in the mire. I would rather see the Forest Service choose the *best* candidate for a job, considering only their skill or knowledge." It is notable that she and her husband, also a retiree from the Tonto National Forest, maintained separate residences when she took her promotion on the Coronado National Forest, until she retired.

She is inspired by the excitement and enthusiasm of new employees coming into the service, the new ideas and technology they bring with them. She derives joy from passing on knowledge to them and giving them the skills and tools to build the future. She said, "To know that we are leaving them places to go where they can discover, learn, become stewards and eventually move into natural resource careers. I love being a part of building that pathway."

Connie poignantly summed up the years and the journey she has been on, every character she has met and how they changed or influenced her life:

> There are so many times I'd love to write about. So many people that have come and gone that not only made a difference to me, but also had an impact on the science of the land. Our Forest Service mission "caring for the land and serving people" encompasses everything I am. It gave me permission to be who I am, to go places I could only dream about and to meet the most interesting people you only read about in storybooks. You find out in your Forest Service journey that they are real! Pick up a book about the first forest rangers and read their life stories, it doesn't get any better than that, especially for a little "Bama" girl who only lived through story books about the old west. Now there are women who serve in our military as soldiers and not just medical fields. Women in the Forest Service have brought so much to the great world of natural resources. I have seen a tremendous change through the years in all aspects of the service. I've seen great female leaders emerge and it was inspiring to have our first female Chief with Gail Kimbell; she led us well. I am proud to be a part of an organization that values us as individuals, that embraces our differences, our knowledge and our unique skills. I am looking forward to a future where the workforce diversity is equal across the board. We are getting there, but still have a way to go.

Chapter 6
Engineers

Engineers in the Forest Service have a wide range of responsibilities. Usually they work out of a supervisor's office at the forest level, and serve all of the districts of a forest. Engineers also work closely with other federal agencies such as the Bureau of Land Management and Federal Highway Administration, state and local governments, tribal governments and many private cooperators.

The transportation program of the Forest Service provides for construction, maintenance and preservation of the national forest road system, which provides access needed to meet the agency's plans and goals for management, public use and protection of National Forest System lands. Recreationists use over ninety percent of the road system. The program also provides engineering support for road construction and maintenance activities on commercial timber sales and land stewardship contracts.[34]

Forest service engineers ensure the safety of bridges and other structures on or along the roads and trails within the national forests. They approve the design of structures and do safety inspections.[35] They engineer and manage thousands of dams located on national forests that are either owned by the agency or are authorized by special use permits.[36]

The Forest Service houses the Geospatial Service and Technology Center (GSTC) and the Remote Sensing Applications Center in Salt Lake City, Utah to provide geographic information products and related technical and training services to Forests.[37]

Two Forest Service Technology and Development (T&D) centers are located in San Dimas, California, and Missoula, Montana. Detached Washington Office employees with expertise in areas such as civil, mechanical, aerospace, electronic and logging engineering, recreation, forestry and social science staff the centers. Equipment specialists, engineering technicians and publications and other personnel also support the T&D centers. In addition, some project work is accomplished through Forest Service employees on other units, contracts and agreements with other agencies. Their objective is to improve business practices and keep employees safe while working in a forest environment. The program uses technology to solve resource management problems of national significance in partnership with other federal, state and international agencies, and with private industry, academia and consultants.[38]

Management of more than 20,000 pieces of fleet equipment falls to the engineering department. The fleet includes automobiles, trucks, construction equipment, boats, snowmobiles, motorcycles, trailers and equipment for fire fighting, law enforcement, agriculture and other special needs.[39]

Facilities engineers manage buildings such as offices, warehouses, research laboratories, visitor centers, houses, dormitories and shower houses. The Forest Service builds a few buildings each year, using renewable energy technology whenever feasible. Several of their buildings are on the National Register of Historic Places. The Forest Service is committed to engineering that incorporates sustainable design features.[40]

The two women whose stories are featured in this chapter are leaders in helping the Forest Service identify sustainability issues and goals and to match its environmental footprint to its stated goals.

∼

Sarah Baker, Engineer

Sarah Baker (figures 94 and 95) hails "mostly" from New Orleans, Louisiana. Her maternal grandmother lived in the country in southwest Louisiana, where she raised cattle. Sarah was able to spend much of the summers and many holidays with her and with her paternal grandparents, who lived in a small town nearby. She had the best of both city and country worlds.

Outdoor Women inside the Forest Service

Figure 94. Sarah Baker sports her uniform — on her first Forest Service job, Kremmling, Colorado.

Figure 95. Sarah Kremmling, Colorado — chopping wood for winter heat.

She grew up on horseback at her grandmother's and riding her bicycle with her friends. She spent every possible moment outdoors, and she enjoyed football, baseball and other sports. She said she was much more interested in football than in dolls. She grew up close to her male cousins, although she had female friends. She characterized herself and most of the girls she was close to when she was young as tomboys.

She loved science and math, and did not do as well in English and social studies, although she mostly made As in school. She was able to go to a small private school from seventh grade to graduation, where she was offered advanced science and math classes, as well as more interesting English and social science classes.

Sarah grew up in a close-knit family. Her father was a social worker, who ran a home for kids with emotional problems and learning disorders, and her mother was a teacher. She had a sister nine years her junior. Her family camped, mostly in the Smokey Mountains, and spent a lot of time together outdoors. Her father died when she was fourteen, and after that the family traveled more, backpacking in Alaska, spending a summer in Europe, and camping all over the United States and Canada. They spent a lot of time hiking in the Rockies. Sarah started both cross-country and downhill skiing as an adult, in Girdwood, Alaska, near Anchorage.

She fell into her profession accidentally. She attended Emory University in Atlanta, Georgia for a year, then dropped out and ran off to Alaska, to the tiny Portage Valley on the Kenai Peninsula, where there were "eleven people and twenty-seven dogs." She worked there for a few months and then moved to Girdwood, Alaska where she worked at low-paying jobs for several years. Those jobs included ski lift operator (that's where she learned to ski), liquor store cashier and flagger for a wrecker service located at that gas station, general contractor/day laborer and waitress "for a crazy lady who owned the restaurant." She says she had a wonderful time, but realized she wanted to make a living, be able to buy a house someday and "generally not have to worry about surviving from one paycheck to the next." She found out that the only people hired off the streets by oil companies in Alaska were engineers, so she decided to be one. She earned bachelors and masters degrees in engineering from The University of Texas.

Sarah said, "Fortunately, I love engineering. What can be better than applied mathematics? And physics? Now I'm in a position where I have a job I've more or less created, working to help the Forest Service become a real conservation agency by adopting more sustainable operations." In 2005, while she was working in the Southwestern Regional Office in Albuquerque as the water/wastewater engineer, the Forest Service tackled implementing an Environmental Management System (EMS). Sarah attended meetings and learned the EMS "language," and was chosen in 2007 as one of four people to put the program together. Sustainable Operations was starting up at the same time, and the region created the position of Sustainable Operations Coordinator and EMS Program Manager, which she filled. In about 2010, the Planning Rule litigation slowed the national push for EMS. Her main focus then became sustainable operations, trying to help make the agency more efficient and aware of resource consumption (Figure 96).

EMS was still a requirement for the agency for its sustainable operations requirements. A Climate Change Action Plan and Scorecard was being created by the Forest Service, which at first did not include sustainable operations. Sarah and her colleague, the executive director of sustainable operations western collective, got sustainable operations into the agency's Climate Change framework. Her goal was to implement EMS, a continuous improvement process, for her region's Climate Change Action Plan.

Sarah said, "This process gave me experience breaking down barriers among specialists who were not accustomed to communicating with each other, and coordinating contacts. I've detailed into projects for the deputy chief of business operations

Figure 96. Sarah (second row, 3rd from left) with the Sustainable Operations group.

Figure 97. Sarah, "Best Days" — Santa Fe National Forest Wilderness.

in the Washington D.C. office even though I don't normally work for that part of the agency. I get to work with enthusiastic, passionate people from all over the country and throughout the agency, and feel like I'm leaving behind a good legacy. I couldn't have dreamed anything better than I have."

Sarah has held a succession of different jobs, leading to her current position. During college the U.S. Geological Survey (USGS) sponsored her as a research assistant. Her first engineering job after college, in 1987, was with the Water Resources Division of the USGS in Jackson, Mississippi.

The USGS did not encourage mobility, and she wanted to travel to other places. In 1989, she was hired as an engineer for the Routt National Forest in Colorado, and from 1991–1995, she worked as the facilities engineer on the Chugach National Forest in Alaska. She next worked as the facilities engineer in Santa Fe, New Mexico (Figure 97).

As a facilities engineer, her duties were dealing with building problems, facilities construction and overseeing contracts. She did mapping and surveying. As a surveyor, the work usually took her to a building or recreation site on a road, but she

Figure 98. Sarah checking map in her Forest Service rig — pre-GPS.

sometimes had to travel cross-country, and needed to be proficient with a map and compass (Figure 98).

In Alaska she carried bear spray, but not a gun, which is required of employees who work in more remote sites. She had a couple of encounters with bears, but nothing dangerous that required her to use the bear spray. One time she was sitting in her car in a parking lot watching the moon, and a bear looked into her car window.

Illness eventually led Sarah to apply for work in the regional office in Albuquerque, New Mexico, where she worked until retirement in 2017. Her days were spent on conference calls, editing or writing papers, making presentations within and outside of the agency, meeting with groups of sustainable operations workers, giving advice on technical aspects of "green" engineering design and hooking up people to work together (Figure 99).

Although each day was somewhat similar, she addressed issues that were completely different. She feels fortunate to have worked at the national level on many projects after arriving in New Mexico, and thoroughly enjoyed details to the Washington Office.

Engineers

Sarah figures she probably turned deliberately to a non-traditional job. She said, "I certainly didn't turn away from it. I grew up a tomboy, and was always drawn to activities that were dominated by males. When I started engineering school in 1981, I was definitely a minority. When I started graduate school in engineering in 1985, I was the only female in the water resources department of the civil engineering school. For the first few jobs I had in Forest Service engineering, I was either the only woman, or one of the few." She never felt excluded because of her gender. She said, "I was used to being one of the only females, so I felt at home with my male engineering coworkers."

Figure 99. Sarah teaching a leadership class.

She spent many years as a forest facilities engineer in Alaska and then New Mexico. She said she dealt with discrimination off and on, and remembers being referred to as the "girl engineer" when she was in her thirties.

She enjoyed the hard work including contract administration and repairing facilities, and she always managed to prove herself. She said, "It always seemed to take my male coworkers in the field by surprise that I knew what I was doing and enjoyed getting in the trenches (literally)."

She said she experienced fewer problems in the Forest Service than she had in earlier government and civilian jobs. She stayed with the Forest Service because, she said, "I really enjoyed being out in the forests. I did — and still do — love my job."

The most difficult situation Sarah has faced so far in the Forest Service was unrelated to sex discrimination. It was supervising a woman who spent two years fighting breast cancer before she died. She and Sarah worked very closely together, and had a rocky relationship at times, but once she was diagnosed, both of their lives changed. Of her employee and colleague, Sarah said, "She'd always been a dedicated employee, and taught me a lot about many aspects of my job. Her Forest Service job was what she used to define herself. When she was facing her final days, she continued to work, but we spent many hours talking about life in general, and she told me she regretted not understanding earlier that there was more to life than work. Now that I'm battling breast cancer, I think about her often, and the memories are probably more poignant."

Sarah faced another special challenge that she shares with younger employees at every opportunity. As a new facilities engineer in Alaska, she worked herself into the ground. For the first few years, she enjoyed the fast pace and the stress. Then she hit a wall and made herself sick. Because of the stress she had created, she had to leave Alaska and start over in New Mexico to keep from literally dying on the job. Unfortunately, once she got to New Mexico, she discovered that she had developed lupus, most likely from being under such constant

Figure 100. Sarah in Russia — sister program in Siberia.

stress for years. That situation limited the amount of time she could spend outdoors, which she truly missed, "especially in this magnificently beautiful state." She encourages other employees not to make the same mistake.

Sarah's greatest passions are learning and sharing knowledge with others. She hopes she can help others understand the importance of protecting our environment and learning to live within our ecological means.

Sarah shared some of her most extraordinary work experiences. About twenty years ago, she was chosen to go to Siberia with the Chugach National Forest's forest supervisor and public affairs officer on the first trip for the Sister Forest Program (Figure 100).

Figure 101. Sarah (3rd from right) — Forestry Commission of the UK in Scotland.

The ten days included "the road trip from hell" where seven forest representatives traveled for three days in a tiny jeep on a road either choking on dust or, when they stopped, covered with mosquitos; "the boat trip from hell" where they traveled the open ocean in a tugboat with one possibly-imaginary arctic suit and water in the diesel fuel; and "the helicopter trip from hell" where they almost collided with another helicopter coming out of a cloud. But she said it was an amazing trip that she completely enjoyed. The people were wonderful, and the scenery was spectacular.

In 2011, Sarah went to Scotland to meet with the United Kingdom Forestry Commission (Figure 101). She had been working virtually with their EMS coordinator. Since she was the last remaining regional coordinator working on EMS, she was sent to Scotland to share experiences with their Forestry Commission, and then hosted their coordinator on a visit to the United States. They have since done several international presentations together virtually, and plan on more meetings in the future to share sustainable operations experiences. Sarah is also hoping to work more closely with the EMS coordinator for forestry in British Columbia. She said, "But Scotland always has been one of my favorite places, so I was incredibly lucky to have the agency send me there."

Sarah has detailed into the Washington Office several times for short stints. She loved working on national issues. On her first day on the first detail, however, someone asked her something that started with, "What would you do if . . . ?" She said, "I gave that person my opinion about the subject, and less than twenty-four hours later my opinion was national policy! I learned quickly to be careful about what I said to anyone."

Anna Jones-Crabtree, Engineer

Anna (Figure 102) was born in Colorado Springs, Colorado, and grew up there on twelve acres that her family still owns. Colorado Springs was a lot more rural at the time than it is now. The family moved to Ridgefield, Washington, also a rural area where Anna's parents still live, when she was eleven. She lived there through high school.

Figure 102. Anna Jones-Crabtree at Buck Mule Falls.

As a child Anna loved the outdoors. Her family spent a lot of time camping and hiking. They spent many summers in her mom's native Indiana.

Anna was a 4-H member. She liked math and science, played the oboe and loved reading. She is the oldest of five children, and was always responsible, the one to set an example. Always an achiever, she was Valedictorian of her high school class. She attributes her success to having had a very supportive family.

Anna's parents had the idea of sustainability before it was mainstream; they started soil building in Colorado on a garden spot, and practiced water conservation. They had plans to build a solar home in Colorado before they moved to Washington.

Anna says she always knew she would be an engineer. Her grandfather, father and all her siblings are engineers. In sixth grade she thought she would be a botanist. She said, "Both have come to fruition in some form; I am an engineer and I own a farm."

She deliberately chose a non-traditional career. In college only about fifteen percent of her engineering classmates were women. She saw engineering as a pursuit that would allow her to have a steady income and support her other interests. She said, "It was something challenging that few others pursued. Looking back, I seem to choose the nontraditional route often."

Anna received her undergraduate and master's degrees from Purdue University in Indiana. Her undergraduate degree is in construction engineering and management. She liked managing and piecing things together. Her master's degree is in civil engineering with an environmental orientation and emphasis in construction.

She was a teacher's assistant in the materials lab at Purdue, working for a professor who was also the co-op education professor. He had a contact with the Forest Service in Colorado, which led Anna early in 1992, to her first permanent Forest Service job on the Grand Mesa, Uncompahgre and Gunnison National Forest headquartered in Delta, Colorado. She spent seven years there, promoting from GS-9 to GS-11 in-place.

The majority of her work was outdoors, especially in summer, overseeing contracts. In the off-season she prepared specifications and designs for upcoming projects. She supported the facilities program.

Early in Anna's Forest Service career she enjoyed projects such as campground rehabilitation and restoration. Later she relished opportunities for more non-traditional engineering, offering leadership in sustainability. She said, "I love being able to put the pieces together to solve problems and see how things fit within a larger system. I am a visual learner and like tangible results. Rather than designing infrastructure such as buildings I'm now designing organizational solutions through collaboration, partnerships and supporting behavior changes."

In 1996, Anna took a leave of absence to attend Georgia Tech University for her PhD. It was the only school in the country at the time with an engineering curriculum around sustainability. She was able to continue working for the Forest Service, half time at her request, at the Region 8 office in Atlanta while attending Georgia Tech. She attended two quarters, and then went back to Colorado for summer work, on a campground rehabilitation project.

While at Georgia Tech, because of her background and knowledge in sustainability, Anna was assigned to a Forest Service team by the national landscape architect, developing The Built Environment Image Guide. The guide was to develop architectural design guidelines that included sustainability. Anna was bothered by a dichotomy between the Forest Service being a "conservation" agency and its actual footprint when it came to consumption and natural resource management. She wanted the Forest Service buildings and operations to embody sustainability. This was also why she had gone back to school. She said, "I wanted to find a way to create more alignment between the agency mission and our actions, and what I saw coming in the future." Her role on the team was to integrate the sustainable building ideas into the guide.

Anna finished her course work for her PhD after two more quarters in Atlanta, and returned to work in Colorado. In 1999, she was hired as the forest engineer on the Big Horn National Forest in Wyoming. She completed the dissertation for her PhD remotely from Colorado and Wyoming in 2001. She was on the Bighorn until 2006. During that time she continued to model sustainable operations efforts. Through a partnership with the Department of Defense and Department of Energy, they installed the first stationary fuel cell in the agency at a ranger station. It is no longer there as the technology was not yet up to the task. Anna said, "I was lucky to establish a pilot effort between Region 2 and the Washington Office."

Eventually Region 2 established the first sustainable operations coordinator position in the nation, which Anna filled. The position was advertised as a virtual duty location, and Anna was able to move to Helena, Montana where her husband, Doug, had been living since 2001. He worked for the State Department of Agriculture, and they had been a dual-career couple. The reason Anna was able to move to Montana (Region 1) when it was a Region 2 position was because the regions were already collaborating on sustainability efforts across regions, and Region 1 also contributed some funding to the position. In 2009, the Western Collective for Sustainable Operations was formalized.

Up until late 2015, Anna held the position of executive director of the Sustainable Operations Collective that she was largely responsible for establishing. This collaboration recognized that field based approaches and doing things together rather than each unit separately would help reap the benefits of sustainable operations more quickly and with less duplication of effort. The position was responsible for supporting collaboration on sustainable operations across all of the Forest Service regions in the west (national forests in Washington, Oregon, California, Nevada, Arizona, New Mexico, Utah, Idaho, Montana, North and South Dakota, Wyoming, Colorado and Alaska) and the Rocky Mountain Research Station, to further the adoption of sustainable operations practices. She provided leadership to reduce the agency's environmental footprint by implementing alternative fuels, waste prevention and recycling, environmentally friendly purchasing, energy and water conservation and renewable energy projects. She grew the program to a nationwide effort in service to the agency as a whole. She said, "This supports even greater return on investment for the Forest Service for this work." She was able to use a detail opportunity into the business operations deputy chief's staff to support greater coordination of this work with the national office in Washington D.C.

Anna also served as the founding co-chair of the Greater Yellowstone Coordinating Committee's Sustainable Operations Subcommittee, which engaged six national forests, two national parks and two national wildlife refuges in seeking collaborative sustainable operation opportunities throughout the Greater Yellowstone Ecosystem. She was also lead author for two national environmental footprint reports for the Forest Service.

Anna's day-to-day world involved engaging a lot of people, "to coordinate and champion conservation and institutionalization of a more robust consumption ethic." She spent lots of time on the phone, conference calls and lots of computer time. She did a lot of telecommuting work, which allowed her to work from home about three days per week and accommodated coordination among the far-flung locations that she served.

Anna enjoyed working with so many different people, hearing their success stories. Her position worked for the deputy regional foresters who comprised the board of directors for the Sustainable Operations Collective. She said "I supported them in modeling good behavior demonstrating a consumption ethic, and they engaged in helping to figure out how best to move this work forward so it became more of our agency culture. I ensured we were transparent and clear on what we could deliver with the resources and capacity we had available and that those outcomes were meaningful to the field." Because of her track record, leadership provided significant support to sustainable operations efforts and her ability to empower innovation across the agency was significant and unique.

Anna is another of the women profiled in this book whose work is not just a career, but also a lifestyle. Her passion for sustainability extends to her personal life. She and her husband, Doug, run a large organic farm, Vilicus, where they raise over twenty crops and model sustainable processes (figures 103 and 104).

Doug farms full-time there, and during growing season, Anna splits her time between farming and her full-time Forest Service job.

Anna said, "While I am proud of the changes I have fostered within the Forest Service I am realizing that my next work will be to find a way to spend more time focusing on sustainability across our food system, and developing more sustainable farming systems. My husband Doug and I are attempting to do this with our farm, using it as a model and an experiment. Our farm has given me many opportunities to advocate for greater sustainability in our farm/food systems. My greatest goal is to foster sustainability on the planet. If we can shift our food and farming system we'll have made great progress, or at least progress that gives me hope for the future."

Figure 103. Anna in rye field at Vilicus.

Figure 104. Anna in flax field at Vilicus.

Anna counts as her number one most extraordinary experience committing to their farm as an adventure. Juggling a full time job and the farm takes everything she has. She said, "It is both taxing and fulfilling, emotionally, physically and intellectually."

She ranks as her second most extraordinary experience actually jumping in and buying the farm, actually commit to buying and starting a farm from scratch.

Another extraordinary experience for Anna was being one of the Donella Meadows Sustainability Institute's Fellows. She was one of twenty students from across the globe in her Fellows class, all engaged in some sort of sustainability work.

Anna has received well-deserved recognition for her work, and of that she said the following:

> I am gratified that my lifelong efforts have received recognition, that I've made a difference. In 2006, under my leadership, the Sustainable Operations effort for the Rocky Mountain Region received honorable mention in the White House Closing the Circle Awards. In 2007, I received the Environmental Protection Agency's Environmental Achievement Award. In 2009, as part of a Ranger Station Extreme Makeover team, I won the White House Closing the Circle, Sowing

the Seeds award. In 2010, I won the White House Sustainability Hero award in recognition of significant environmental contributions to a federal agency. In 2012, the Sustainable Operations Western Collective received the Chief's Award for Leading in the Business Environment as acknowledgement of our collective efforts across the west to not only reduce our environmental footprint but our cost footprint as well.

Though Anna has appreciated her years with the Forest Service, it has come with big challenges. She was sometimes impatient because the agency still had a long way to go to integrate a consumption ethic as part of its day-to-day stewardship work, and then she hit a huge speed bump. Due to some unfortunate organizational shifts in late 2015, Anna's years of endeavor began to unravel. After about a year of watching the decline of the organization she had put heart and soul into, she came to question the true commitment of the Forest Service to sustainability. She felt the program received lip service, but leadership had eroded. She saw no support for inspiring leaders from the highest levels of management. She likes to be positive, but felt abandoned. The turn of events left her questioning how committed the Forest Service is to taking care of their people to the extent that she has seriously considered resigning. She does realize that much of this is political, and admits that perhaps her program had moved forward at a pace that the agency was not prepared for. She still holds hope that eventually the pendulum will swing back in her favor. She said, "There continue to be many challenging opportunities to change the system so these ideals are supported. It's important to surround yourself with people who understand and hold a similar vision, and keep building the skills needed to be your own best champion. I hope for positive actions to continue at the ground level. This experience has reinforced for me that the best work really happens at the local level."

Whatever decision Anna ultimately makes about her future, the world is bound to benefit, and there is no denying the Forest Service is a better place for her legacy.

Chapter 7
Unique in Some Way

Many programs within the Forest Service are or might be unique and vary from the normal way the agency's business is done or how it is funded.

The profiles of workers presented here are representative of this uniqueness and did not fit neatly into any of the other chapters. Overall, the work varies and may or may not be different than that of other positions. This may be because it is part of an experiment such as enterprise teams of regular agency employees with special authorities that make it easier to hire extra help and to work from virtual offices. It may be different because it is a training program such as Job Corps. Technical and professional employees from different levels in the agency are represented here. Also presented in this chapter are examples of jobs and programs from the three branches of the Forest Service besides the National Forest System (NFS) — State and Private Forestry, Research and International Programs.

Entomologist Iral Ragenovich works for State and Private Forestry — Forest Health Protection. Hers is an example of a job that serves the NFS but does not receive funding or direction from NFS. Entomologists serve all of the NFS and they work on all of the national forests. They are located in a forest health protection regional office or in service centers or zones that may sometimes be located in a forest supervisor's office. Their mission is to survey and monitor insect and disease occurrence and to provide technical assistance and training to land managers. For instance, they do an annual aerial detection survey of all forested lands in order to monitor and record trends and occurrence of damage. They provide training to land managers on how to identify insects and diseases and provide recommendations on how to manage them. They also do technology development and transfer by interpreting and explaining findings from the Forest Service research branch to facilitate application of the science in the field.

Insects and diseases, as with fire, do not recognize boundaries, so to be effective state and federal agencies need to work together. Forest Service entomologists monitor insect occurrence and provide technical assistance to forestland managers on all federally managed lands, working with BLM, NPS, USFWS, BIA and DOD (Department of Defense), and they monitor and provide insect and disease assistance to state and private land managers, mostly through grants and agreements with state forestry agencies. Providing assistance to all of the national forests as well as the other federal land managers and state and private entities offers opportunities for broad experience.[41]

Research has been a part of the Forest Service mission since the agency began in 1905, with the aim of employing the latest science to improve the health and use of the nation's forests and grasslands.[42] The work of research scientists today focuses on providing knowledge and tools to managers and decision-makers to assist in the sustainable management and use of natural resources.

Connie Millar is one of about 500 Forest Service researchers who work in a range of biological, physical and social science fields to promote sustainable management of the nation's diverse forests and rangelands. She is a senior scientist from the Forest Service Research Branch at the Pacific Southwest Research Station located in Albany, California. Scientists like Connie who work out of regional research stations, conduct research targeted to the needs of the geographic region served by their station.

To ensure that research applies to all the diverse natural resources across our country, Forest Service research is conducted nationally at over sixty-seven laboratories, organized around seven regional research stations plus the International Institute of Tropical Forestry in Puerto Rico and the Forest Products Laboratory in Madison, Wisconsin. A network of eighty-one experimental forests and ranges complements the work of the research stations, and Forest Service research collaborates on forest research with other countries.

The Forest Service takes a global approach to natural resource management through its International Programs division.[43] The program fosters sustainable natural resource management, biodiversity conservation, climate change mitigation and adaptation and disaster preparedness and response throughout the world. The Forest Service supports American forestry interests while exposing the agency's workforce to new ideas and experiences, which creates opportunities for agency employees in the United States to perform work details in other countries. Forest Service staff from many fields, including wildlife biologists, forest economists, hydrologists, disaster and fire management specialists and policy makers work with partners overseas to address the world's most critical forestry issues and concerns.

These include reduction of illegal logging and wildlife trafficking, climate change and development of environmental projects that connect communities with nature locally. The program takes the United States public land management expertise for protected areas to other parts of the world, which includes natural resource based tourism. Watershed rehabilitation, green infrastructure, human health and environmental education for youth, families and school groups are other areas of concern. The program encourages collaboration around the world to support, strengthen and enhance habitat management for and conservation of migratory species.

International Programs promotes research, monitoring and management of areas worldwide that are experiencing insect, disease and invasive plant disturbances. Partners work together to develop strategies to prevent, suppress and control outbreaks, with an emphasis on links between changing climate and pest conditions. Work includes partnerships to integrate remote sensing and field measurement technologies to monitor the health and status of forests, and to apply these technologies to specific management issues. The program also advances the professional capacity to avoid catastrophic wildland fires in partner countries, and better manage fires that do occur.

The broad-based objectives of International Programs provide opportunities for employees from any discipline to participate. Sarah Baker, Stephanie Connolly, Cheryl Hazlitt and Traute Parrie, women portrayed in this book, have been involved in International Programs, and have taken their areas of expertise to places like Russia, Turkey, Hungary, Egypt, the West Bank and Africa. Garrit Craig's inclusion in this chapter is multi-purpose, since she worked for years as a smokejumper and then in the Disaster Assistance Support Program (DASP) area of International Programs, and recently made another career move into a deputy district ranger position.

The DASP was created to provide the U.S. Agency for International Development, Office of U.S. Foreign Disaster Assistance (USAID/OFDA) with essential technical support in disaster response management, planning, operations, preparedness, and prevention. It is a cooperative program between USAID/OFDA and the Forest Service International Programs. The DASP enhances and supports OFDA's capacity to respond to disasters by mobilizing more than 250 disaster management experts from the Forest Service and other domestic agencies.

The USAID/OFDA coordinates responses to dozens of foreign disasters each year and recognized that the Forest Service has emergency response capability and experience in wildland fire fighting that is applicable to any type of disaster situation. Because the Forest Service developed and implements the U.S. National Incident Management System to systematically manage fires, a partnership was created in order to build and improve USAID/OFDA's disaster response capabilities, applying that system.

The role of the DASP has expanded to include emergency support functions such as training, developing and improving USAID/OFDA's methodologies for disaster response and coordinating USAID and Embassy disaster preparedness. The DASP consists of 16 full-time staff providing Response

Management Systems services in policies and procedures, competency-based position management, organizational learning, training, capacity building and mission disaster preparedness. Program elements include formal training including disaster simulations and team building exercises, consultations to develop procedures and protocols and study visits and seminars to observe field level disaster operations. When Forest Service or other agency's staff detail to DASP, they may serve in many of these functional areas.

Smokejumping is a distinct fire fighting program. The Forest Service runs several smokejumper bases, in the western states.[44] Smokejumpers are a national resource that travels from the bases to locations all over the country to fight fire in remote areas as needed. Smokejumpers are required to meet rigorous physical fitness criteria that only the most motivated women have been able to attain.

Law enforcement employees are different because they provide services to ranger districts and forest and regional offices, but are in a separate division, the U.S. Forest Service Law Enforcement & Investigations unit (LEI) which is headquartered in Washington, D.C. Each region is headed by a Special Agent in Charge, who oversees law enforcement personnel within the region.[45] Law enforcement personnel may be stationed on ranger districts but do not report directly to district rangers.

Nine women are profiled in this chapter, including one posthumous account of a law enforcement officer who gave her life on the job.

~

Cindy Champion, Assistant Smokejumper Foreman

Cindy Champion's childhood home in San Anselmo, Marin County, California is twenty minutes from the Golden Gate Bridge and San Francisco. With her mother, father and two older sisters, Cindy (Figure 105) flourished growing up there, and frequently returns to this "beautiful area with lots of nature around." The mix of oaks, redwoods and civilization

Figure 105. Cindy Champion and Smokey.

still inspire her to seek exotic places where nature thrives and diversity abounds.

As a child she was very active, practicing ballet beginning at age five, whenever she wasn't biking around the neighborhood, playing in the creek, building forts and camping in the backyard. Soon she bounced into gymnastics and dove into synchronized swimming. Her mother was an active swimmer, and steered Cindy and her sisters into aquatic sports. Competitive sports kept her interest from grade school to high school. As a youth, she was a member of a local baton-twirling/marching corps organization that competed around California, Washington and Colorado. As she grew older her preference evolved to individual sports, running, mountain biking and skiing.

While in third grade, Cindy was rated at the fourth-grade level in math and the fifth-grade level in spelling. Her teachers wanted to advance her a grade, but her parents resisted, believing Cindy would become disconnected to friends her own age. An avid reader, she developed a love of language and words, and also became intrigued with faraway places, which eventually lead to a geography degree in college.

Cindy's love for the outdoors stemmed from her family's interests. Together they went on many vacations to national parks, as well as skiing at Lake Tahoe. The family also belonged to the Napa Valley Ranch Club, where Cindy enjoyed swimming, horseback riding, shuffleboard and camping.

She knew she wanted a career where she could work outside. Right after high school, she held some indoor jobs such as accounting, insurance, a city planning department and restaurant work, and she wanted something more physically demanding and outdoors. After graduating college in 1991, with a degree in geography from Humboldt State University in northern California, Cindy worked at a couple of temporary positions, including a summer at Samuel P. Taylor State Park as park aide. In 1993, she returned to Humboldt State and studied botany. While there, a botany classmate told her about work on the Winema National Forest. She landed a position on their plant crew in 1994. After arriving on the Winema she found out that she was going to fire guard school. In those days, all new hires on the Winema were sent to fire guard school to become qualified as wildland fire fighters. She didn't particularly enjoy guard school, and thought she had found out enough about what fire was about, until she went on her first fire assignment and became hooked on fire.

On one of her early fire assignments she learned about hotshot crews, and was intrigued by the challenge of working with a team of fire fighters that are heavily relied upon by the rest of the fire organization. She worked on a Sequoia National Park fire crew in 1995, and in 1996, she became an Arrowhead Hotshot (Figure 106).

Cindy did not really know what she was getting into when she embarked on a fire career. She just knew she loved being in a job that paid her to physically work out, and opened doors to the outside. To this day she still loves the travel that comes with each new fire assignment and the chance to experience a variety of places.

Figure 106. Cindy (second from left) and Gal Pals — Arrowhead Hotshot reunion.

She worked with Arrowhead again in 1997, and then took a season off and attended massage school. The next year, influenced by a friend on the hotshot crew, she became interested in smokejumping. She applied to most of the Forest Service bases in the country and had offers from two. She had, in 1992, before any fire experience, gone skydiving on her own, "just to find out how it felt." She also took a friend skydiving for her birthday. With that prior jump experience she knew what to expect, but remembers being apprehensive when she started smokejumping. Even after seventeen seasons of jumping fires she still gets a little nervous when she's standing in the door of the plane about to exit from 1500 feet above the ground; she copes by keeping focused on the jump, something she has now done over 275 times (Figure 107).

Fire fighting is a seasonal occupation, usually running from May to October. Cindy was routinely employed by the Forest

Unique in Some Way

Service at least part-time during most off seasons, and in some years has worked for a Yellowstone snowcoach company as a winter guide, taking visitors on ski tours into Yellowstone National Park.

Work at the smokejumper base starts at 9:00 a.m. and usually ends at 6:00 p.m., except when there is a fire assignment; then all bets are off. A normal day consists of roll call, followed by a briefing on the current fire situation, weather and the day's work assignments. Then there is an hour or more of physical training (PT); smokejumping is arduous duty and requires all the jumpers to be in top physical shape. PT consists of running, weightlifting, pull-ups, push-ups and sometimes biking.

After the morning briefing, Cindy makes sure she is ready for a fire assignment if she is on the list for the first plane load.

Figure 107. Cindy in the sky.

A load consists of eight smokejumpers — that is how many people the airplane can carry at once. Being ready means that her gear is jump ready to go out the door of the plane, along with her personal gear bag containing everything she needs for the fire — clothes, food and sleeping gear. If it is rainy season it includes rain gear, and mosquito repellent if they are heading to Alaska.

Weekly proficiency jumps keep smokejumper skills current. The practice jumps can take most of a morning, and after the jump, the training continues with a debriefing, checking chutes and critiquing the jump video of each jumper's exit and landing. In the afternoon following practice jumps, those who jumped re-pack their chutes.

On fire assignments, the day starts earlier, usually by 6:00 a.m., with the crew receiving a briefing and instructions. They normally work all day and into the evening, typically sixteen hours, scouting, building hand line, cutting saw line and hazard trees, establishing hose lays, burning out, patrolling and mopping up the fire.

Working on a fire is physically challenging, so Cindy and her coworkers pace themselves. At the end of each fire shift, they feel "a good tired," and sleep well. They usually eat meals together, breaking for lunch and dinner, or sometimes just "eating a bean on the fly." If they are near a fire camp that is providing meals to fire personnel, they will eat at the camp. Otherwise, they each eat whatever they brought in their fireboxes. After two or three days they may be eating army-type provisions known as MRE's (meals ready to eat). If a fresh food supply is delivered (air-dropped via helicopter or paracargo), they share the "manna from heaven."

Cindy did not deliberately set out to have a career as a wildland fire fighter, but happened into it by chance. Growing up in the 1970s, near San Francisco, a time and place of significant social change, coupled with "my being independent and somewhat rebellious," led her down the non-traditional path.

In San Anselmo, Cindy was far removed from the fire world, barely knew there were jobs fighting forest fires. Along with Cindy, only one other woman was on the 20-person hotshot crew in 1996–1997. There were just a few women smokejumping when Cindy started. In the 1980s and 1990s, the Redmond jump base had about eight women out of a total of fifty jumpers; in 2001 and 2002, there were only four. In 2018, there are still only four. In 2018, of approximately 445 smokejumpers nationwide, about twenty were women. My opinion is that it is a particularly unique woman who has both the will and the ability to meet the physical demands of smokejumping. Minimum smokejumper physical standards include being able to execute seven pull-ups or chin-ups, forty-five sit-ups, twenty-five push-ups, a 1½-mile run in eleven minutes or less, carrying a 110-pound pack for three miles in ninety minutes or less (pack-out) and passing the standard fire fighting pack test for Forest Service.[46]

It is palpable that some males in leadership have had issues with women fighting fire, for a myriad of reasons, which resulted in women not always being considered for developmental opportunities. Sometimes there is a subtle undercurrent of attitude from peers (other than immediate peers she works closely with — "the bros") and supervisors against women in leadership positions, leading Cindy to believe she has been held back. Part of this may be because she is somewhat introverted, quiet and reserved. It may also be things happen the way they do for reasons beyond anyone's capability of knowing. Cindy said she truly believes that, "I'm right where I'm supposed to be."

Being very physically fit, Cindy has had no difficulties performing the demanding tasks of fire fighting. She exercises at work, and in her own time religiously works out at skate skiing, classic skiing, telemark skiing, ski touring and biathlon (skate skiing and shooting), hiking, running and biking. She also does yoga to maintain flexibility and stave off stress and strain injuries.

Cindy experienced one serious injury on a fire jump on the Umatilla National Forest. Her parachute collapsed at about forty feet above ground, her speed increased, and she fell about thirty-five feet, slightly missing the jump spot. She blacked out at about twenty-five feet, and does not remember landing. As a result her body was so relaxed at impact it probably saved her from worse injury. Badly bruised and banged up, she had no broken bones. When she came to, her crew who had watched her fall had found her and called the medivac helicopter, and she was flown to the hospital in Pendleton. The result was she was off the jump list, on light duty, for six weeks.

The scariest experience at the top of Cindy's list of extraordinary experiences is being chased by a grizzly bear. She was with an eight-member crew from the lower forty-eight in Alaska in 2000, her second year of jumping. The crew was on a cabin protection assignment, setting up sprinklers and burning out around four cabins at Wilderness Lake.

After the work was done, some of the fire fighters were fishing to supplement the MREs they were eating. One of them caught a fish as a bear came out of the woods, and the fish was reluctantly left to the bear. After the bear ate the fish, a problem ensued; the bear would not leave, and there were no more fish. The crew started a Mark III pump and chain saws to make noise, but the noise did not scare the bear away. They retrieved a rifle from one of the cabins and shot it into the air to scare the bear, but it still did not leave. Chris, one of the crew, called dispatch, who replied all they could do was send a helicopter. Eventually, the bear did wander away on its own accord, without mishap.

Cindy and one other person separated from the crew and headed back to camp. In camp all alone, Cindy was about to remove her radio harness, when she looked up and saw a different bear appear, only ten to fifteen feet away from her. Her thumb just happened to be on her radio mike, and she instinctively radioed her partner to announce the presence of the adjacent bear in the form of a scream. Backing away from the bear, fear took over, and she did what she knew she should not do during a bear encounter, and chose flight over fight. One of the crew who overheard the scream on the radio started running toward her. He then tripped on a rock hidden behind the tall grass and broke his knee, decommissioning him and his jump career.

Cindy ran through some tiny spruce trees toward the lake. It was about fifty feet to the shore. The rest of the crew was on its way back to camp, just as the helicopter was arriving. Cindy's crewmember, Ray Rubio, was at the lakeshore, and saw her coming, with the bear chasing her. He yelled at her to stop running. She stopped, turned, put her arms up, trying to look big and started yelling at the bear. The bear did not see Ray, and Ray threw his radio at the bear and stood next to Cindy, which did phase the bear. Ray and Cindy backed up gradually toward the lake, and soon were in the water. The helicopter was approaching. The bear remained undaunted. The helicopter pilot forced the blades down to blow air onto the bear to create some space. He landed the helicopter on the lakeshore between the bear and Cindy and Ray standing in the lake, and they were able to scramble into the helicopter to safety.

After a few helicopter circles in the air the bear departed, and Cindy and Ray were placed back down on the ground a few hundred yards from camp. The crew immediately broke camp and got flown out. After communicating with the incident commander, it turned out that the crew was stationed in a densely populated bear area. The next day a dead moose was found on the lakeshore, with five bears feeding on it.

Outside of work, up until recently, Cindy led a rather solitary life, and lived alone for a long time. Working in fire and being away from home most of the summer is not conducive to stable long-term relationships. In 2014, she reconnected with an old friend who also works in fire. The friendship grew into a steady relationship and she found new joy being with someone who loves her for who she is, and is understanding of her work as a smokejumper.

After getting her permanent job in West Yellowstone she lived in Bozeman, Montana, two hours from the jump base. From 2005 through 2008, she wintered in Bozeman and worked summers in West Yellowstone. In 2010, she bought a one-acre lot in West Yellowstone and in 2014, began building her dream home. In 2016, she completed building an energy efficient house, complete with yoga room and solar panels.

Cindy feels incredibly lucky going on fires and being able to experience phenomenal places, such as Alaska and the Grand Tetons. Fall is her favorite time of year to be in the woods, when the aspen leaves are turning. She said, "I know fire season is almost over and it is the last time to embrace it. It's so great to be out in the woods in the fall, in perfect temperatures and surrounded by color, reflecting on the season past."

Cindy's greatest goal is to continue learning, to evolve to be the best person she can be, to "be who I am meant to be." She is dedicated to nurturing herself, eating healthy food and daily physical exercise. She strives to have her own life together so she can "be a role model to help others reach their full potential."

~

Garrit Craig, Deputy District Ranger and former
Disaster Program Specialist,
International Programs

Garrit Craig (Figure 108) was born in 1980, in Salem, Oregon and grew up in the small seaport town of Newport, Oregon, where her dad has lived since he was in high school. He is a skillful carpenter, and a bit of a perfectionist, a trait passed on to Garrit.

Garrit's mom, now retired, was a first-grade teacher for many years. When Garrit was eleven, her parents divorced, and her mom moved to Eugene, Oregon. Garrit stayed with her dad in Newport. Her older brother lived with both parents on and off, then ended up finding a place with friends. Today he is an accomplished portrait artist, who teaches art in Eugene. He is an important part of Garrit's life, and she sees him often.

Garrit sympathized with her dad, and that influenced her decision to stay with him. She also realized more freedom living with him. Not that she could run wild; her father was disciplined and structured, and taught her how to work hard.

Figure 108. Garrit Craig on fire assignment in Alaska.

She said, "He made me tie knots, do the dishes, stack wood, wax the truck and finish a project to the best of my ability." He did not necessarily know how to raise a teenage daughter, but he was an adventurous outdoorsman, and always pushed Garrit's boundaries. He took her camping, fishing, swimming and exploring. He took her and her girlfriends camping and boating, and loved to make a campfire and tell a good story. This remarkable story illustrates her dad's influence on her relationship with the outdoors and adventure:

> When I was seven or eight years old my dad drove my brother and me to Depot Bay, Oregon. The city of Depot Bay was planning to take down a well-known giant spruce tree, and he wanted us to pay our respects. The tree was on the edge of a cliff, seventy feet straight up from the ocean. Sixty-five feet tall, with a trunk about six feet in diameter, it was filled top to bottom with sturdy, jagged branches that had been twisted and strengthened over time from the consistent force of the northwest wind.

There was a storm brewing in the area, and the wind was blowing hard. Standing underneath the spruce staring up at hundreds of branches, it looked like I was standing underneath an enormous Christmas tree. None of the branches had been trimmed, and the tree looked wild and untamed.

My dad grabbed a branch shoulder high and starting climbing the tree. My brother (four years older than me) started right behind him. Before I knew it, they were lost between the branches, and I was left to follow. I started up, one branch after the other, but the wind kept scaring me from ascending quickly. I could hear my brother and father screaming to each other between the wind and branches. I couldn't tell what they were saying but the distance in their voices lead me to believe that they were ascending quickly. Their progress both scared and excited me. Somehow my brother got ahead of my dad and I heard him yell he was almost at the top. I made it about ⅓ of the way up, and couldn't move any further. The branches started to be out of my reach, and the distance to the ground was too far. My brother sat up there for a while, feeling the wind swaying him from side to side. Slowly, they both came down, and when they were in sight, I started to descend. As we reached the bottom the rain started, and we got in the car and left. My brother wouldn't stop bragging about how wonderful it was at the top, and I couldn't have been more jealous.

Garrit was an adventurous leader of her group of girlfriends. She said, "I was always keen to make a bonfire on the beach, find a hidden campground and explore any undiscovered area. When I was eleven or twelve I used to ride my bike down to the beach or reservoir to find an adventure that was often just beyond my comfort zone."

She was contentious with boys and frequently argued with boys in her classes. She said, "High school, in particular, felt like a never-ending war. The boys that I battled often outnumbered me and my girlfriends would just roll their eyes. I remember one girlfriend saying, 'Why don't you give this one to them Garrit?' I wasn't aware that I was starting most battles — I just saw something I didn't like and challenged them. Most of the time, I felt defeated in the end. I was convinced that all the boys just hated me, when it was probably more that they felt challenged."

Garrit maintained a relationship with her mom, which improved, as Garrit got older. Her life choices reflect the influence of both parents. After teaching first grade for fifteen years, her mom taught in alternative high schools and in the Eugene prison. She helped both at-risk youth and incarcerated adults earn further education, and provided informal counseling services to them. She now works in her garden and is involved in the Eugene Jung Society and other community organizations.

Garrit got good grades in school, gravitating toward English. She got her undergraduate degree in English literature from the University of Oregon in Eugene, wanting to learn to write better.

She had taken up smoking cigarettes when she was fifteen or sixteen, experimenting with her friends. At eighteen she decided she did not want to be a smoker when she went to college, and she started running in order to quit. She was surprised that it worked, and she found that it served as a tool for her to funnel her energy. She said, "I fell in love with running and the adrenaline rush it gave me and started seeking out further sports and adventures that would challenge me physically and expose me to the many offerings in the Pacific Northwest." She already was a strong swimmer and had swum throughout her youth on the swim team, and she is an avid surfer.

At twenty-one, Garrit joined her first fire crew, a contract crew out of Eugene. She had worked as a waitress for a couple of summers, and that "wasn't cutting it." She had been thinking that fire fighting sounded cool. An old boyfriend who knew she was interested led her to this crew, and at the beginning of the summer she got her first fire call. She said, "It's not the route I would recommend for a new fire fighter because money is never guaranteed and it's often a ragged group of folks — but this was the route I took."

Contract fire crews are used by fire fighting agencies as needed, augmenting regular agency fire fighters, so assignments can be sporadic. Crewmembers are not paid for their fire fighter training time, or time traveling to fires, and their wages may be lower than those of fire fighters hired directly by the Forest Service. The alternative path to a fire fighting job with the Forest Service is to apply for a seasonal job that provides regular work all season (between fires they do work such as building, campground and trail maintenance), and benefits such as being paid during basic fire fighter training.

About half of Garrit's contract crew had been to jail or was coming off of drugs. A quarter of the crew were older guys who had been doing this since they were kids, and a quarter were "kids" like Garrit, looking to make money for school. There were two girls, and around eighteen men. Garrit offered a colorful description of her experience:

> It was a typical Eugene show — half hippy, half red neck, half whatever. Almost everyone smoked or chewed, and weren't in the best of shape. The crew boss had blown out both of his knees, and the assistant smoked two packs a day. I was surprised to find that I was among the top in-shape. But it still wasn't easy. I had to learn to develop a sense of brute strength beyond the comfort boundaries of an everyday athlete. I had to stay dirty, hike with unbearable blisters, carry heavy, awkward weight and stay up through the night. During the day we slept in the middle of fire camp in a busy, hot, sun exposed field. My first assignment, we were placed on night shift, monitoring and gridding the area.

Gridding is a systematic process in which the crew lines up to cover burnt ground after the fire has been contained and mopped up, to ensure the fire is out cold. On smaller fires the crew is on their hands and knees, sifting the ashes, touching literally every inch of burned area. On large fires, not every inch is touched, but it is still a tedious process, as Garrit related:

> The most amusing part of this assignment was the grids. Gridding is something that all fire fighters despise. The process is a mock on the human intelligence, but is somehow scrutinized as never being performed successfully. On this particular night we couldn't keep our line straight — mostly out of defiance and lack of will. The crew boss and the squad boss started nipping at each other early in the night as one

would inch ahead of the other. An hour or two into the grid they were full-on ready to fight, throwing down their helmets and gloves, and cursing so loud that the entire crew could hear. Another squad boss stopped it before they engaged. Hats and gloves went back on, and they returned to their positions. It wasn't discussed later — it was just a normal day's event.

Garrit went out on four fire fighting tours that summer. She never felt threatened or unwanted as a woman. She felt accepted and respected as a hard worker. She said, "There were a few skuzzy individuals who would look at me in a certain way or speak disrespectfully towards me. I was a fighter, and would challenge their remarks. Much like in high school, I would challenge anything thrown my way. And as a twenty-one year old, I didn't know how to pick my battles, often leaving me in an uncomfortable position." Overall, she got along well with the majority of the crewmembers. She loved to laugh and listen, and she made friends with people that were outside of her norm. She said, "The experience gave me a wider perspective, challenged me physically and exposed me to individuals and to country that I wouldn't have otherwise known."

Garrit spent ten years working in the rough-and-tumble world of fire (Figure 109). She does not think she deliberately sought non-traditional work; she was just looking for adventure, and liked the hard work. People who knew her were not surprised by it.

Her second season she felt she was ready for the next step in fire so she applied for various hotshot crews throughout the Northwest. One out of ten crews called her back. The supervisor phoned her and wanted to know her exercise routine and why she was interested in being on a hotshot crew. A few weeks later she was hired.

She skipped out of college early that year, and arrived at the base in a rural Northwest town around the middle of May. She was one of two women on the crew, and the two were immediately split up into different modules and pressured to compete with one another. The other woman was rarely there

Figure 109. Garrit (right) — rough and tumble life of fire fighting.

as she had college courses to finish, and missed most of the assignments that Garrit went on. It was only the second year of the crew's existence, and the supervisor and assistant were young and inexperienced leaders.

At first Garrit kept up on the runs and the hikes, coming in at about the middle of the pack. She had come to the crew in good shape, but still needed to work on hiking with speed. She was not mentally ready for what the summer held, but she was determined to make it work. She reminisced about her first difficult challenge on the hotshot crew:

> About a month into the crew I met the boys at a local bar for some drinks. The senior lead on the crew, and the most physically fit and respected among the members, approached me in front of the boys and told me specifically what he would like to do to me. The other boys sat around listening and said nothing. I wasn't afraid and fought him with words of pride and disgust. Eventually, however, after receiving enough of his remarks, my tears started flowing.

One of the boys tried to comfort me, telling me this was a test that I had just passed. I didn't understand any of it and I walked home broken. When I got home I told my roommate in secrecy what had just happened. She was much older than I and took it upon herself to tell the district fire management officer (FMO). I would have preferred her to keep her secrecy, and wasn't aware that she said anything until the crew met for a meeting mid-morning the next day. With eighteen men standing there, and one female, the supervisor explained that an anonymous member on the crew had been sexually harassed by another crewmember, and that the FMO was now watching the crew closely. All eyes turned to me. It was the start of my decline on the crew.

After that, Garrit's physical abilities began to decline. Having once made most of the hikes with the crew, she started coming in last. Most stopped speaking to her, including the supervisor and the assistant, and dismissed everything she said as stupid.

While the crew waited for their first call, the supervisor and assistant continued to prep the crew through long practice hikes. In one instance the supervisor showed disrespect and a lack of dignity for all the members on his crew, not just women. Garrit said, "On one particular hike, half of the crew had picked up a virus of some kind. Half way up the hill, one of the lead swampers in dire need pulled down his pants and started shitting in place. As the crew hiked around him, the crew boss yelled, 'Pinch it off and let's go!' The lead swamper pulled up his pants, and continued shitting as he hiked the rest of the way. Once we started going on rolls, the supervisor and the assistant made it clear that we weren't to talk to other crews, and they often disregarded the direction of the division in terms of safety and orders."

One man on the crew who was Garrit's buggy mate sent her signs of support from a distance. After some of her rougher days when she felt the most isolated, he would secretly smile and pass her gummy bears. He encouraged her on hikes and supported her in any way that he could. He was careful not to be seen by the other men. Garrit said, "But his effort meant a lot to me at the time and helped me to keep trying."

The constant harassment was demoralizing. Three and a half months into the season, she had lost fifteen pounds and was much weaker than when she arrived. She had been placed in the back of the line with the rake — a common place for the one female on the crew. During their third tour the 'lead saw' on the crew took her aside to tell her how much he disliked her, even the sound of her voice. She tried to fight back, but eventually dissolved into tears. She said, "This, surprisingly, hurt me more than the previous harassment I had received because it wasn't a test, it was pure hatred and it was becoming infectious. Every day was a battle, and every day I would lose."

Garrit's fifth tour that season was pivotal. During her last hike on the crew, the crew boss had to stay behind with her as she struggled up the hill. She said, "Snot was pouring out of my nose, and I was stumbling over my feet. The crew boss said very little to me. The next day I was stopped by the crew boss and assistant at the top of the hill and told I had been washed from the crew and was officially a safety hazard. I was flown back to the base, and drove home to Oregon a few days later. The supervisor told me that I wasn't hotshot material, and at that point, I believed him."

At home Garrit started to regain her strength, and she had time to reflect on her experience. She said, "Looking back now, I see that I could have picked my battles more wisely, talked a little less, listened a little more. I don't know if I was hotshot material, but I know I came to the crew physically prepared, and the isolation and lack of support broke me down. I'm not sure that the supervisor and assistant ever wanted me to succeed. As I left, the supervisor said, 'We keep trying to hire women, and they keep failing.' As though women just didn't belong on a hotshot crew — especially their crew. And the men on the crew were all young twenty-somethings who were following leadership and trying to make it themselves."

Garrit was heartbroken, but she had a girlfriend who was a smokejumper who encouraged her to keep going in fire. They started running, hiking and lifting weights together, and her confidence returned. She applied for several heli-rappeller bases the following summer. A base in eastern Oregon called her references, and got a bad report from her former supervisor, who said, "She wasn't tough enough for our crew." Garrit explained her experience, and the assistant respected her for her honesty and strong will.

Two months later she was hired as a rappeller. Rappellers are used during initial attack (IA) on fires. They are dropped with ropes over the skids of helicopters, allowing them to get into tight spots. They also may support IA by landing, then support crews, bucket and supply drops and reconnaissance.

One of her first challenges as a rappeller was hiking a five-mile course with eighty-five pounds on her back. She remained in the middle if not the top of the pack for the entire course. In the middle of the training period, the assistant supervisor approached her and said, "Your old hot shot supervisor doesn't know shit! We're glad to have you!" Later that summer Garrit learned that her old hot shot crew was disbanded for the season for getting into a bar fight where one of the crew bit off

Figure 110. Garrit (left) and "fellow" rappellers pose during a pack out after a fire.

a man's ear, as well as a chunk of meat from his chest. The offender was one of the senior leaders on the crew.

During her two years as a rappeller, she received a lot of support from her peers and supervisors. She said, "Sure, I still had battles, but I wasn't outnumbered or isolated. I also made some strong girlfriends that year (Figure 110). We encouraged one another, competed with one another, and formed strong bonds in the process."

The following year she applied to be a smokejumper. Again her old supervisor's reference haunted her, and again she made her case. She was hired for rookie training at the age of twenty-six. Ironically, the nightmare she had survived on the hotshot crew helped her endure the rigorous rookie training. She said:

> I worked my ass off that winter to get into the best shape that I could and showed up in May to take the physical training (PT) test and start rookie training. I passed the PT test fine, but the all-night line dig was where I began to show my weakness. We finally got to lay down around 3:00 or 4:00 am for a couple hours, and then were back up, hiking to the top of the hill, preparing for the eighty-five-pound pack test. I was exhausted at this point and was the last to pick up a pack, leaving me with the most misshapen pack of them all. I remember looking down at it, not having the energy to pick it up, and wondering how I could possible succeed. I was behind from the start of the hike, with the strongest female rookie trainer on my tail, trying to get me to quit. My body started failing me, my feet were falling asleep, and each step was a challenge. I was so far behind that I still don't know to this day where everyone else was, but I kept going, and they couldn't get me to take my pack off. Towards the end of the hike, I had broken open two holes on my lower back, and I was walking straight on a gravel road. Three trainers were beside me, and at this point they knew I wouldn't give up. The whole experience brought me back to my hotshot days, and the thought of being washed and being kicked off the hill kept me going — regardless of what my body was saying.
>
> Through trial and error I made it through rookie training. I was punished often for my errors, and even had to do extra

training at night. I was by no means the star — far from it. But I tried hard, and I passed.

Reporting to her base after training, she soon realized that rookie training was just one of many challenges she would face. She came in with smiles, laughter and a big mouth, and was met with disdain, especially from the younger second or third year men.

One man, a bully and a leader to the younger guys, hated her from the start. He despised her presence on the crew, and working with him was a nightmare. As she described it:

> He would talk shit about me, lie about me, make me work the dirtiest tasks and at one point I even found him pissing towards me — anything he could do to make me feel unworthy and unwanted. I remember jumping a fire where he was in charge, and the fire was entirely out of his control. Additionally, we had an injury, a broken femur, the cargo had landed in the fire, and someone had burnt up a parachute. Meanwhile, the wind was blowing stronger up hill and the fire was growing. He was completely lost as to how to handle it all and a more qualified Incident Commander (IC) was called in to take over. Another woman and I tried putting out the burning cargo with one of our water QB's [five gallon jugs of water in a balloon with a spigot, contained in a box with a handle for a pulaski, all of which are slung over the fire fighter's pack — quite heavy] and he yelled at us. I tried relaying a radio message, and he again yelled. When we got off the fire forty-eight hours later, I wasn't surprised to hear he had spent the previous night bitching about me. I found him alone in the loft the very next day. I was fuming mad, so sick of taking all of his shit and finally I laid it on him. I told him the fire was out of his control, that there were a lot of things that went wrong on that fire, and none of them were my fault or his. I further told him that in the future if he had something to say to me to stand up like a man and say it, rather than pussyfooting behind my back. And lastly, I asked him to please not put his insecurities on me. These may not sound like strong words, but without his squad of followers surrounding him, the man had no backbone from which to respond. His head

was low, and he had nothing. Everyone heard this conversation from a distance, and no one came to his aid. I guess they figured it was my time to make a stand, and I felt a sense of honor and accomplishment as I walked away. It was a little piece of victory that I received that year.

Garrit "worked my ass off" as a rookie and was essentially a mule for the season. She carried everything — water, saws, gas, tools, equipment and many other things that never got used. She worked hard, jumped twelve fires, met all physical challenges including several hike-outs carrying 120+ pounds and kept up a fighting spirit. Incredibly, they made her wear a cowbell around the base so they would always know where she was. She took it in good fun, and ran up and down the stairs, hammering the bell until it drove everyone mad and they forced her to take it off. She said, "I played the game as best I could and enjoyed the time I had as a rookie."

Still, she was not well liked by the young men at the base. In time she realized that fitting in had less to do with a good attitude and a strong work ethic than with how the image of a female jumper threatened their image of what they thought a jumper should be. She said, "With time, I can more clearly see that these men were merely trying to figure out how to be men, and I, a female presence, got in the way of that."

While the group of young men on the crew kept their distance, the older guys took her under their wing, and called her "Big G." She said, "I can't say that they lent a helping hand during my battles, but they watched me from a distance and gave me their approval in subtle ways. One guy gave me tips from time to time. During my second year as a jumper he told me, 'Shame only works if you give a shit so don't be sucked into unnecessary work.' Jumpers were always striving to work more, regardless of whether the effort was effective."

Her third year as a jumper, Garrit was hired at a base she could finally call home — West Yellowstone, Montana (or as referred to by locals, "West"). Tucked in the trees next to the tiny West Yellowstone Airport, she lived with a group of jumpers that started and ended the day with laughter and hard work. She said, "West Yellowstone is not only one of the most beautiful jump bases, but it holds some of the friendliest, hard-working, most hilarious group of jumpers around. Finally the culmination of defeat, hard work and struggle that I faced over the years led me to my reward."

The men at West were indifferent to her at first. She slowly grew on them, and carved out a place for herself, becoming well known as "Big G" to both young and old. Her hard work and good attitude was appreciated, and she was respected and encouraged.

Most of the jumpers lived on base at West and every night there was a bonfire, fishing trip or swim in the surrounding lakes and rivers. They would also ride their bikes the two miles to town on dirt roads that lined the airport, and then ride them back, "drunk as hell," laughing and falling the entire way. Garrit said, "Our war wounds the next day (or mine in particular) were a great starting topic for the morning briefing."

During four fire seasons Garrit explored and ran on trails throughout West and the adjacent Yellowstone National Park. When the men at the base learned this, they insisted she carry bear spray. She said, "This turned out to be a wise idea as often I would see grizzlies in the distance, and be grateful that I had my minimal protection."

With more experience as a jumper, and having landed in a good place, Garrit picked her battles wisely, but she still faced challenges. On a boost to Redding as a third-year jumper, she landed on a fire in the Sierras with a mixed load of jumpers from West Yellowstone and Redding, California. The fire was a large Type 3 and she and her jump partner were tasked to watch a burning snag in steep terrain on the edge of the line. As the wind picked up, so did the flames, causing the fire from the snag to jump outside the line. Garrit got on the radio and informed the crew that there was a slop-over and that they needed men and saws quickly. She was nervous in her communication, unsure of the appropriate actions, and very conscious that she was on another crew's turf. While they waited and tried to control the slop-over, she continued to check-in about their request and expressed the need for help. It took a while for back up to arrive, and by the time they did, the fire had carried a good couple of acres. The men all looked surprised by the progression of the slop-over, and with their help they were able to contain it.

That night, sitting around the campfire, the leader amongst the group called Garrit out for not relaying the urgency of the slop-over. She was sitting on a box with six or eight men surrounding her, looking down on her, and blaming her for almost losing the line. She said, "At the time I felt overwhelmed by their criticism, and didn't have the words to respond. My jump partner was speechless, and didn't stand up to defend me." One of the senior women on her crew, who was passive but experienced, stood up and kindly said that she had heard Garrit ask for men and a chain saw several times. She then looked at them with a glance that questioned their judgment. With those words, she defended Garrit when no one else

Figure 111. Garrit — Anaktuvuk Pass, Alaska.

could. Garrit said, "I'm still unclear as to why the crew took so long to respond to my call. They either didn't care to take me seriously; or they just weren't listening. Regardless, it was easier for them to blame me than take responsibility for their lack of action. My nervousness left an open window for their criticism, and this lesson further taught me to speak with authority and confidence regardless of how I felt; something, I believe, comes much easier to men than women."

The woman who defended her became one of Garrit's best girlfriends in fire. Garrit lived with her for three out of her four years at West Yellowstone. Garrit said, "Senior to me, and the only other woman on the crew, she was my mentor and friend, and an essential guide in my growth as a female jumper. Like me, she strived to maintain her femininity, and authentic self — which is no small venture for a female in fire."

During her time at West, Garrit jumped fires in Yellowstone National Park and all over Montana, Idaho, California, Oregon and New Mexico. Her last two years as a jumper she went to Alaska, jumping ten fires. She described that experience in detail:

> My glory jump in Alaska was a jump north of the Brooks Range into Anaktuvuk Pass — an area that the Alaska jumpers rarely see. The locals drove out on their quads to watch us jump and greeted us as heroes (figures 111 and 112). We were shy, but appreciative. Once we hit the ground, paradise quickly turned to hell as we realized that swarms of mosquitos were inching their way into our jump suits (Figure 113). But regardless of the bugs, we all felt privileged and thrilled to have jumped that stretch of land. Surrounded by green valleys, mountains and icy rivers, this jump marked one of the most beautiful areas that I have fought fire (Figure 114).
>
> Three days later, when the jumper plane came to retrieve us in the local village, the lineup of wealthy touring Americans waiting on the tarmac reminded us all of our privileges as jumpers. Watching them load into their first class transport, we all felt proud and honored to serve some of the most beautiful parts of our country. When we got back, the base was filled with jumpers from every base in the United States.

Everyone had heard about our jump, and as we walked into the ready room, steady, playful banter marked our moment in fame. Unable to keep a straight face, I was absolutely beaming with excitement and joy.

During her third year at West Yellowstone, Garrit had started to get involved with USFS International Program's Disaster Assistance Support Program (DASP). During fall and winter of the prior two years she had earned a graduate degree in social work from Humboldt State University in California, and there she heard about the program. She applied for DASP orientation, and was detailed to the program.

In the winter of 2010, she completed a long-term detail with the DASP and was encouraged by the program manager to apply for an upcoming position. After ten years in fire she was ready for a chance to enhance her career outside of fire. Her knees were starting to bother her (a common malady among long-term fire fighters), and she was becoming anxious about being injured on a jump, thus reducing her ability to run, surf,

Figure 112. Garrit (wearing cap, rear center) and fire crew, Anaktuvuk Pass, Alaska — local kids.

Unique in Some Way

Figure 113. Garrit (far right) and crew on fire in Alaska.

Figure 114. Garrit's photo of Anaktuvuk Pass, Alaska.

swim and practice yoga. Her first year as a jumper she had cracked her pubic bone and spent six months without activity, and never wanted to go through that again.

Garrit was also becoming weary of constantly relocating. She had bounced between college housing, working seasonally on fires and living with her mom during college. While earning her graduate degree in social work, she'd also worked in several related social service jobs, including work with abused kids, school and medical social work and as a teacher at alternative schools. During some of her off-seasons from fire she traveled to Mexico or Costa Rico where she rented a house and spent months surfing, then it was back to wherever the next fire season took her. It suited her adventurous spirit, but she was ready to settle down some. She said, "I was growing tired of a bag-a-bound [sic] life style. Over a course of ten years I had moved *thirty* different times, and was ready to settle in one place. The most troubling part of all was that I had finally found my home in fire in West Yellowstone, and sadly, it was time to go." It was hard to say good-bye to her base, and all the wonderful people that she worked with. She did so at their end of the year party. It was the last day that the plane was on contract, and of course, they got a jumper call, thus leaving six to party. Garrit said, "It was a perfect group of some of my favorite jumpers. We all got drunk and swung at piñatas at a bar at one of the near-by lakes. Drunkenly, the boys grouped around me and told me I would be missed. I honestly couldn't have asked for anything more."

She took a new position with the DASP in the fall of 2011, and worked for the Washington, D.C. office for five years. She did not fight fire again after she left, but the identity of a fire fighter is something that she will always carry. She said, "I know now that all the excruciating trials and disappointments in my ten years of fire were the seeds of my true core strength. Each individual who challenged me was a teacher who helped me to find my courage, confidence, pride and determination. I doubt I will ever again find a career with such camaraderie, laughter, hard work and adventure."

By contrast, her DASP work was physically tame. She exercised her brain more than her body. Her work was in a capacity building program, consulting other countries to improve their disaster management functions. She developed partnerships and agreements, and support training in systems management, budget and finance.

For five years Garrit worked in Washington, D.C., then was able to work remotely from Eugene, Oregon when she needed to return there for family reasons. At this writing she had been in a new job for about three months, as deputy district ranger on the Siuslaw National Forest, and has bought a house and moved to Florence, Oregon for the promotion. She is back with the most important people in her life — her dad, mom, brother and girlfriends she has known since middle school. She does not know where her career will take her next. She plans to stay with the Forest Service and continue moving up in the organization. For the moment she is content with the progress she has made and is thrilled to own a dog and have a house where she can hang up her surfboards.

Kristine Fairbanks, Law Enforcement Officer

Kristine Fairbanks (Figure 115) was shot and killed September 20, 2008, while on duty in the Olympic National Forest, south of Sequim, Washington. She was fifty-one years old. I gathered her biography through interviews with her father, her daughter and her best friend.

Kristine (Kris) Fairbanks was born in Sault Ste Marie, Michigan on the 4th of July, 1957, where her father, John Willits, was attending Michigan State University's summer camp. Upon John's graduation, he moved to Washington State for his Masters degree in forestry. After college he worked for Weyerhaeuser as a forester in Chehalis, Washington for five years. A job as Washington State Parks administrator then took the family to Olympia. Travel associated with that job frequently took John through Port Angeles, where he chose to settle. After three years with State Parks he finally got a

Outdoor Women inside the Forest Service

Figure 115. Kristine Fairbanks with her beloved canine.

job offer at Peninsula Community College in Port Angeles as a forestry teacher. He had not taught a day in his life before that, but he quickly developed a program, and spent the rest of his career there. Kris inherited her father's adventurous free spirit, which led her eventually into her too-short career as a Forest Service law enforcement officer.

Along the way, she avidly explored life. Many diversions meant that she went through four colleges in seven years en route to her college degree. Intermissions from college included one year in Colorado as a ski bum, and a bike tour of Europe. She took lots of science classes, and when she decided it was time to finally get her degree, she took a counselor's advice and chose geology. She got the degree then never worked a day as a geologist. Her father says she never even liked rocks.

Kris never considered the mundane when it came to employment. Before working for the Forest Service, she worked for several seasons for the Park Service. Her first Park Service job was in Denali, Alaska. She drove there by herself with two cats. When she came home she worked two winters doing ski patrol at Olympic National Park, and then several seasons as a temporary employee for Olympic National Park at Crescent Lake.

It was no doubt Kris' adventurous spirit that first attracted her to law enforcement. At Crescent Lake her duties included doing traffic stops and campground compliance, and she loved it. She went to the national law enforcement academy in Georgia while working at Lake Crescent. That background led to her permanent law enforcement position of twenty-two years with the Forest Service in Forks, Washington. During some summers she worked on a fire crew for the Forest Service, out of Forks, where she ultimately settled, marrying and raising her daughter, Whitney.

She had met her husband, Brian, while she was working for the Park Service. Brian was also working on a summer job for the Park Service. He was part-time Park Service, and a full-time city cop for the city of Forks.

When asked what Kris had liked as a child, her father said, "Anything contrary to what her dad wanted." He says she was stubborn, a free spirit. When asked what some of her extraordinary experiences were, he said, "She probably wouldn't tell her dad." He knew she was very adventurous. She climbed most peaks in the Olympics when she was young. She attempted Mount Rainier twice, but got altitude sickness before getting to the top. She and her husband had floated a major river in Alaska, navigating alone.

Interestingly, Kris' family did not engage in outdoors activities as a family; her mother did not like camping. They did some boating, which her father characterized as pretty civilized. Kris developed her interest in activities such as hiking, climbing and camping on her own, and did them with friends. She did some duck hunting with her dad. He had a couple of hunting dogs, but Kris had not yet developed what later became her passion for dogs. When she was seventeen she had a job working at the Olympic Game Farm in Sequim, Washington. She once brought a horse home, and another time two wolves. For a short time she thought she wanted to be a veterinarian.

Her keen interest in dogs came later, when she was active with her daughter, Whitney, in the Forks 4-H. She volunteered for the 4-H dog club. She later became a K-9 law enforcement officer and eventually a trainer, for the Forest Service.

Kris' dad does not think she deliberately sought out a non-traditional career. She had an affinity for the outdoors, animals and interacting with people. She just wanted a job, and found a good fit. She had been working part time for the Forest Service when her district ranger told her they had to hire a woman law enforcement officer. Her ranger told her that she got the job, "but don't screw it up."

The forest had never had a law enforcement officer before, male or female, and gave Kris no help as she learned the job. They gave her a truck and sent her off to the woods on her own. Early in the job, Kris discovered that a timber management clique had built an unauthorized hunting cabin on the district, and she told her ranger about it. It turned out that he was part of the group that had built the cabin. He instructed the timber crew to tear it down and build it again where Kris could not find it. Kris went through several years without any support from her district. She had a difficult time being accepted by her male co-workers, but as she gained confidence in her own abilities, she earned respect. Eventually, she became good friends with everyone she worked with.

Kris did have a support network with her Park Service and other city and county agency contacts. Her husband is also an enforcement officer for the Washington State Department of Fish and Wildlife, and an agreement between the Forest Service and the state allowed Brian to accompany Kris on the job at times.

While she was working for the Park Service, there had been a big drug bust at Crescent Lake, in which dogs tore up a car to get at the drugs. Since Kris often worked alone in the Olympic National Forest, she thought it would be a good idea to have a police dog to accompany her. She already had the love of dogs going for her.

K-9 Officers was another program the forest had not previously participated in, and Kris assumed they would not be supportive. She spent thousands of dollars of her own money to pay for her training, used her own time to train and bought her first dog, to prove that the program would work. It turned out to be a major success; Kris eventually trained K-9 law enforcement officers and their dogs, nationwide.

Kris always reached out to others, investing both her time and money. She became a pillar in her community, which extended over the entire north Olympic Peninsula. Even more than five years since her death, The Dungeness Ranch Pet Resort in Sequim, where Kris often boarded dogs, runs a memorial to Kris on their web page. So does a BLOG on the 4-H page, memorializing her leadership of their Happy Tails program. Over 3,000 people from the United States and Canada, including the Governor of the state of Washington, attended Kris' memorial service. The most poignant sentiment comes from her father, who said, "She was just a good kid."

Minde Callis was best friends with Kris for about thirty years, and is the godmother of Kris' daughter Whitney. Minde attended college in Port Angeles, Washington, studying forestry. Kris' father was the forestry teacher, and Minde met Kris one day when Kris was visiting his class. In 1978, Minde got a job with Forest Service fire dispatch, and Kris was on a fire crew. She and Minde both lived at the Forest Service work center in Forks, Washington, with a group of other fire employees, and got to know each other well that season.

When Minde developed back problems in 1982, she left forestry work to work for the Clallam County Court District in Forks, from which she retired nineteen years later. Kris also lived in Forks, and the friendship between her and Minde continued.

Minde couldn't engage in some of Kris's physical activities, such as hiking or climbing, but they were kindred spirits in many ways, and did most everything else together. Kris loved clothes, especially sweaters, and loved shopping. She and Minde would stay overnight at Minde's parents' home in Silverdale, to shop there. Kris and Minde's families also did things together, cross-country skiing when their girls were young, then later downhill at Stevens Pass. The girls still spend a lot of time together.

When asked what Kris liked doing best, Minde agreed with Kris's dad that it was probably anything contrary to his wishes — she had a stubborn streak, was a free spirit.

Kris sometimes confided in Minde, telling stories of the dangers she encountered on her job. She dealt with a lot of groups of Hispanics who harvested salal berries in the woods. She was never directly threatened, but she felt uneasy. Kris partly solved that problem by learning some Spanish so she could communicate better with the groups.

She was actually threatened more by loggers and ATV riders. One incident she related to Minde was about a logger stealing firewood. Kris got a call about possible timber theft, and her husband responded with her. When Kris confronted the logger, he started his chain saw and went after Brian, and he and Brian fought. Brian won, and Kris called the sheriff to make an arrest.

Minde said, "In light of the dangers of both Kris and Brian's jobs, they took care to raise Whitney to be strong. They made sure she knew that something could happen to them. I think that having that background helped Whitney through her mother's death."

Kris' daughter was born in Port Angeles, and raised in Forks. At this writing she was attending college at WSU in Pullman. She was studying criminal justice, following in both her parents' footsteps. She wanted to be a police officer; did not know which agency. Whitney grew up with police dogs, and wants to be a K-9 officer, like her mom.

Kris travelled a lot for work. She worked the winter Olympics one year. Once she worked on the Canadian border, intercepting drugs between Canada and the United States. She spent over a month patrolling on the Mexican border, an experience she did not enjoy; Kris felt sorry for the immigrants.

Whitney felt her mom did a good job of balancing time with her, and she did not have negative feelings about her mom's travel obligations. Kris called home daily when she was away, and attended Whitney's sports events when she was home. Sometimes they were able to make a family vacation of her work travel.

Whitney was not aware of Kristine facing discrimination or harassment for being a woman in her profession. She was pretty close with the men she worked with, including those from other agencies. One of her mentors was a Port Angeles police officer, Kevin Miller, who trained her first dog.

Kristine shared thoughts about the dangers of her job with Whitney. She said that is why it was nice to have a dog there. Whitney did not worry about her mom.

Whitney knows she will be faced with same types of situations when she becomes a law enforcement officer. Because both her parents were in law enforcement she feels she is more aware of situations than the typical person. That she aspires to the same profession bears testimony to the good example her mom set for her. Like her mother, Whitney is strong, adventurous, and independent.

Cheryl Hazlitt, Interpretive Planner

Cheryl (Figure 116) grew up in Oklahoma, moving numerous times during her childhood. Her father was a Methodist minister and was transferred to a new church about every two years. Cheryl's family life revolved around church life, caring for her five younger siblings and family activities. Family outings included swimming in local lakes, cookouts and camping, church services and church life. Cheryl always gravitated toward the outdoors. She loved climbing trees, playing in the woods and swimming. The family did a lot of camping when she was young at what is now the Chickasaw National Recreation Area. She has great memories of swimming in the icy cold water that came from underground springs, sitting around the campfire, attending ranger led programs and "my first up close introduction to snakes!"

With six children in Cheryl's family and the limited income of a minister, she learned early to be a "frugal gatherer." Her

Figure 116. Cheryl Hazlitt at work — Flaming Gorge Recreation Area.

mother took the children every summer to pick blackberries, strawberries, potatoes and green beans. They shucked bushels of corn, pitted cherries and peaches and had a full freezer at the end of every summer. Cheryl learned to be resourceful and strategic in planning ahead.

Early in life, she connected to nature as a place to escape the chaos of her big family, and as a spiritual retreat. When she was thirteen, her mother began treatment for cancer. She was ill for nine years before she died, affecting Cheryl deeply. During that time, the family struggled. Nature was a place of comfort and solace. Cheryl found a retreat in a nearby field with a trickle of a stream, wild onions and long tall grasses. It was near the railroad tracks, and she listened to the roar of the trains, wondered where they were headed and was on the lookout for hobos who rode the trains in those days.

She said she would always carry a deep sorrow in all the things she and her mother missed with each other. Yet, her mother had told her when she was very young to be able to take care of herself and not to think she could just rely on someone else. Cheryl strongly internalized this message. She knows she could do it alone if she had to. She missed many normal activities and parenting in her teen years, and the guidance of a mother, but also had no role model for what she could or could not do. She missed out on learning to write thank you notes, dance properly and clean house, but developed a strong sense of intuition and ability to find common ground with others.

While Cheryl was mostly an outdoors kid, she also loved playing house, and had dolls, including Barbies. She liked to create performances with her siblings or classmates. While reading and writing stories came easily, she says, "Math always made my palms sweat." Accordingly, she did not like accounting or bookkeeping, or math word problems. She was often in school plays and performances, because she sang well and music was fun and came easily. Science, biology and outdoor education have always been great loves of hers.

Cheryl graduated from Oklahoma State University in 1982, with a degree in Home Economics Education. During the 1980s, she spent several summers working at the YMCA of the Rockies, in Estes Park, Colorado, where she first realized there were jobs working in the outdoors. She moved to Salida, Colorado for a teaching job, and taught home economics there for several years, working summers in Rocky Mountain National Park as a ranger. She took some time for travel to Nepal and then moved to Fort Collins to go to graduate school in a hybrid program combining natural resource courses with education classes. She has called Fort Collins home since then. Staying in one place to raise a family was a high priority for her, given her experiences growing up.

Thoughts of what attracted Cheryl to her profession bring a wistful smile. She said:

> It wasn't what I had planned my life to be — not at all. I really thought my life would go down a different path, a little more like a simplified Martha Stewart type. The high school I graduated from had forty seniors, about half male,

half female. Of those twenty young women, half of them were either married, getting married or pregnant! I knew I wanted more than just that in a small farm town in rural Oklahoma. When I was twenty-two, graduating from college just weeks after my mother's death, I wanted to escape to something else. Colorado felt like a place I could be myself, with all my emotions, and no pretenses. The mountains didn't care about my sadness, or the mess of my family; they just stood like silent sentinels, witness to my grief, ever present, never judging. I found a great deal of healing in nature that first year, and something else. The processes that create life, our world and each environment are captivating — I still feel that way. That different path has led me to this place where I feel like the luckiest person in the world to be able to combine my love for nature with being able to share the stories and the significance with others, and to share this magnificent planet's systems of life with them.

While in college in Oklahoma, Cheryl had worked for one season as a housekeeper, which strengthened her resolve to get a college degree. After that she got the summer job working in the youth adventure program at the YMCA of the Rockies. Once she started her job as a teacher in Salida, her summer work in Rocky Mountain National Park was first as a front line interpreter (naturalist) for the National Park Service giving an average of twenty-two programs a week, and later as a backcountry ranger. She reflected on this, saying, "Working in interpretation feeds my intellectual curiosity. I begin new projects in diverse places, and meet new people who care deeply about the landscape they live in. I get to discover something new and different. Seeing the passion that others feel about the place they live in creates a sense of humility and admiration, sometimes awe about the earth and the beings on it."

In her fourth year as a teacher, she took a four-month leave of absence from her teaching job to travel to Nepal. Part of her trip was a cultural study through the Colorado Department of Education to meet her teacher certification requirements. Upon her return, Cheryl resigned from her teaching job to earn her graduate degree, pursuing natural resources in a more direct manner. After graduation, Cheryl took a position with USDA/ Boulder County Extension Office as a 4-H Youth Agent. Wanting something more tied to natural resource agencies, she accepted a term (limited time appointment) assignment with the Forest Service. Her first assignment was to develop an interpretive plan for the Beartooth Scenic Byway on the Shoshone and Custer Gallatin National Forests.

Cheryl's current title is Interpretive Planner. These are rare positions in the Forest Service, or in most natural resource agencies today. An interpretive planner assists resource managers in recreation management. They define issues related to human use and experiences, identify and create methods to convey stories and issues to their publics. These issues are translated into design solutions that may create a new site, product or message for the public that assists with management issues and visitor experiences (Figure 117).

Figure 117. Cheryl (right) with graphic designer Donna at installation of Red Canyon Visitor Center exhibits.

Figure 118. Cheryl in Siberia with assistant director of forest reserves for Republic of Kakassia, Siberia.

Interpretive planning is often associated with recreation planning, engineering and landscape architecture. Her group is assigned within engineering because of the interdisciplinary nature of design work.

Interpretive planners have to like people, be good communicators, creative thinkers, quick on their feet, problem solvers and innovative. Cheryl's days vary widely. Her work group is project based, and they can be juggling from eight to fifteen projects at a time. Projects can range from designing five-sign panels, overseeing a visitor center exhibit design contract, writing a forest interpretive plan, or developing a brochure, website or video production.

Cheryl believes her team members are some of the best in the agency. Their combined strength as a team has resulted in doing work in every region and nearly seventy-five percent of the forests. Sometimes they work on international assignments

Unique in Some Way

to assist other countries with interpretative work. Recently Cheryl traveled with Forest Service International Programs (IP) to tour some of the reserves set up in the Republic of Khakassia in Siberia, Russia that are developing tourism sites and training non profit groups that are developing tourism materials. Her group did interpretive planning and exhibit design workshops with the Russians (Figure 118).

She has enjoyed other special assignments with International Programs, working on the coast of the Red Sea in Egypt, visiting a pristine beach with sea tortoise nesting sites, traveling into remote desert country, or meeting the local native tribal women. She worked with the Palestinian Wildlife Society on developing areas that include bird monitoring stations, interpretation and conservation education within the West

Figure 119. Cheryl, West Bank holding a bag with remarkable Palestinian embroidery.

Figure 120. Cheryl (in vest) meeting stakeholders on Volcanic Legacy Scenic Byway.

Figure 121. Cheryl site planning for a stop on a wildlife refuge along the Volcanic Legacy Scenic Byway.

Bank. They visited five different sites, did an analysis of conditions and made recommendations for training, capacity building and future actions that IP could work with the Society to accomplish. She marveled, "I have gotten to train West Bank wildlife specialists in interpretive planning for sites dating back over 2,000 years!"

Those trips require intense immersion in the work project. She works sometimes for ten to twelve days straight to create concepts and ideas for visitor experiences. The projects are hard work providing few breaks. They are culturally profound experiences that she looks back on fondly, years later (Figure 119).

While these assignments represent a small percentage of her career, she said, "They are diamonds in my career, things I could not have gotten to do in any other way. Even for tourists traveling the world, the experience is different, as many of these sites are not open to American travelers going abroad."

There are days when Cheryl drives a scenic byway as part of a planning process and gets to experience the vast mountains, rivers and forests of Colorado, deserts of Utah or volcanoes of northern California (figures 120 and 121).

Occasionally, she works on national-scope assignments for the Washington Office, that can vary from evaluating a proposal for a new visitor center to meeting a recreation or information need with a very short turn around.

These are balanced by days when she spends the entire day haggling on the phone, doing serial conference calls, or organizing mounds of data into something logical. She sits in front of her computer, praying for creative inspiration; trying to stay focused. She spends a good deal of her time managing project budgets, developing a scope of work and putting teams together to complete projects, or problem solving within a project. Interpersonal skills, diplomacy and knowing when to lead and when to step back are important skills.

Cheryl did not have a clear intention of entering a non-traditional profession; she did what she enjoyed and what came naturally. She thinks she grew into feminism without realizing it. Only when someone asked her if she was a "bra burner" did she consider that she was a feminist in many ways, when she replied, "No, that would be a waste of a good bra, but I expect to get paid the same as any man doing the same amount of work." In her first natural resource job with the Park Service, she was surrounded by white males, most in their thirties and forties, and some with traditional attitudes towards women, especially young and inexperienced women.

Her first permanent position with the Forest Service, in 1992, was in engineering. The interpretive planning staff worked with the architect, landscape architects and engineers to design solutions to resource and recreation problems. Cheryl's training in education and interpretation, combined with her master's degree and experience within a federal agency prepared her for this type of creative design work.

When she first began working as a GS-7 temporary, in the Region 2 office in Denver, there were approximately thirty-five people within the engineering office. Of those, there were three professional women, not including the administrative staff, Cheryl and her supervisor, who oversaw the design group, and one female engineer. She walked into the lunchroom once, and the conversation stopped. She suddenly realized they were all men sitting there and they were uncomfortable about her hearing their conversation. She thought, "How should I handle this? What should I do?" Her habit was to try and put people at ease, but in this case, she knew they were uncomfortable because she had overheard a joke with sexual innuendos in reference to a book written with a great deal of sexual content about a woman. It was a defining moment, where she felt the difference between the sexes keenly. She jokingly asked if she was interrupting their book group, and said she was glad to know this was an intellectual group. They all laughed, and Cheryl never experienced any further awkward moments with them. She decided that, "Emotional intelligence is critical for a positive work environment, and using humor and good will is always a better tactic then anger and accusations."

Generally she has always enjoyed her male and female colleagues, and felt welcomed by them, but believes gender differences combined with personalities always come into play in the workplace. As a young seasonal, she had been approached by a much older highly respected man and managed to casually sidestep him, and never be left alone with him after that. When she was younger, she used her wits to get out of these kinds of situations, and now as an older woman, she said, "It sets a slow burn that so often those behaviors were hidden or overlooked in the work place, and young woman (and men) today must contend with them." She thinks that teaching children about boundaries for appropriate behaviors is where it begins.

Issues of pay equity and lack of promotions were more pronounced earlier in Cheryl's career. Yet still today, she thinks that as families struggle with the need for two working parents, one parent, often the woman, typically makes career sacrifices. She has struggled to maintain family and work balances her entire career. She does not fault others for situations, but has felt the sting, when early on, a male colleague received an award for outstanding uniform appearance and she knew that was because his spouse shined his shoes, ironed his shirts and polished his badge — he just had to show up.

Cheryl said, "We didn't have relatives nearby to help us when we were working, parenting and managing a household. There were nights when my kids were little that I would wake in the middle of the night, in a panic. I learned later that is a symptom of extreme exhaustion. But I didn't have a lot of choices, I had to keep going."

She sees many employees dealing with the same issue. Some of them have children with special needs that require extra parenting and supervision. They chose to work for the Forest Service because it fits their values and ethics for land conservation, and is family friendly. She believes the Forest Service has made great gains. She has always had supportive supervisors and a great deal of flexibility in work schedules, but the desire to have a successful professional life combined with parenting creates a lot of inner conflicts for her. She hates hearing, "You can't have it all," and has fought against that her entire adult life. She does think, "You can't have it all, all at the same time."

As for herself, Cheryl has always felt she did not quite fit the mold of a Forest Service employee. She was not trained in forestry, engineering, wildlife biology, or any of the standard disciplines the agency has hired in the past.

She does feel fortunate to have a position within the agency through the years as budgets waned. Her current work group is a hybrid of regularly employed staff and an enterprise team. She considers this both a blessing and a curse. The team is specifically project-funded, like an enterprise team, and operates similar to the private sector, but is held to federal hiring rules, unlike fully authorized enterprise teams.

She thrives on solving complex problems. Work meets her need for intellectual stimulation and creative problem solving. She is passionate about being engaged deeply in life; experiencing it, feeling connected. Expounding on that sense, she said:

> I am most deeply moved by the miracle and abundance of life forces, and the ability to adapt, evolve and flourish. To witness life that springs forth in the face of adversity, be it a tree seedling that emerges from a rock after a fire; a mountainside where growth regenerates after an avalanche; a community that rebuilds after a flood, or individuals who pick up and start again after tragedy. All of these are life that recovers in biological, emotional and creative ways — for me there is a deep spirituality in observing the quiet, enduring patience of nature.

She believes it is her human responsibility to leave the world a better place because of having been here. She said, "How that looks can ebb and flow at times." At the moment, doing everything she possibly can to provide a solid future for her children, so they can flourish, is paramount to her. She said, "That is my primary job for this moment in time."

Considering her greatest goal caused Cheryl a lot of emotion. She has always been rather goal driven, but she can't point to one great thing and say, "This is it." As a child, she expected a higher power to speak directly to her and tell the purpose and goal for her existence. She has grown more pragmatic over the years and sees goals as part of the "fluid river of life's journey." In her twenties, her goal was to travel to Asia, trek in Nepal, and climb some of the high peaks there. She earned a graduate degree, which was another goal. She struggled to have a family and grow a career, which eventually came into being. Now through her forties and into her early fifties, she thinks the goal is to enjoy the here and now, the present life she has, people she loves and work that she does.

She still considers moving up the Forest Service career ladder beyond and feels she is not done within her professional life, and is invigorated by the changes happening in communications, media development and the sheer number of changes going on today.

Cheryl shares her life with an avid outdoorsman who loves hunting and fishing, camping and kayaking. They were introduced shortly after having both traveled to Nepal, on different trips. They have been together for over twenty-five years, through difficult times in careers, family and parenting. They now have two children in their teen years. Cheryl is reminded of her role as a mother in the modeling she does for her daughter, and is caught off guard at times by how much her children need her in these years. She said, "The loss of a parent in a child's life takes a lifetime to figure out, far into adulthood." Her children continue to create extraordinary experiences for Cheryl. She said, "As they grow, mature, learn, become who they are meant to be is an ongoing adventure — that experience is a continuous one!" It pleases her that both her children see her as active and working outside of the home, doing stimulating and fun work. That is the same thing she wants for each of them in their adult lives.

Cheryl hopes to continue to travel and experience other cultures, "to see more of this incredible planet and the diversity of plants, animals and cultures that thrive on it."

Work with the Forest Service International Programs (IP) created some extraordinary travel experiences for her. Traveling to

Egypt and the Red Sea coastline to Wadi El Gemel National Park gave her close-up encounters with Egyptian forms of resource management, and Arabic culture. She worked with the local park staff training them in interpretation skills; exhibit planning and designing visitor experiences. The Red Sea is known for its rich marine diversity caused by the periodic massive flooding of the wadi's (delta's), and during the off time, snorkeling became a fast addiction for Cheryl. She paddled among coral reefs that had every color, hue, texture and shape possible; fish of every color, coral bursting with plant life so unlike any terrestrial life. These waters are what draw huge tourism from Europe to Egypt, and management problems that go with it. She described perfect blue water, white sand and an "entire world underwater that I could only visit, but not a human habitat." She would stay in the water until she was chilled and starving, or had to get back into work more on her assignments.

Another IP trip took her the West Bank and Palestine in 2010, as part of an evaluation team. This effort included assessing several natural and cultural attractions, identifying capacity needs for training and resources, providing training on recreation and interpretive planning with local resources managers and meeting with government officials. She saw a close-up view of the Israeli-Palestinian conflict, understood the Palestinian perspectives, wishes and hopes. She got a close up view of what war and conflict does to natural resources. Giant holes were left by tactile war maneuvers; the rumble of armored tanks and the struggle for a moderate government to maintain control damaged ancient ruins. She described it as "humbling and exasperating, and made me think it must have been similar to our own United States history of removing native people from lands desired by others. Land and people are inextricably linked to each other."

A different assignment harkened back to this experience when Cheryl spent eighteen months working with the Wyoming State Historic Preservation Office (WSHPO) in a partnership with the Forest Service and the Southern Arapaho and Eastern Shoshone tribal groups to write an interpretive plan detailing history of these groups and their collective experiences on the Wind River (Shoshone) Reservation. Cheryl was able to spend time with individuals in their homes interviewing them about their memories, stories and histories. She received opinions and stories that were very different and contradicting to some of the "stories of the west" she had grown up hearing. She said, "Truth comes in various forms and not always with the same capital t."

Preceding these travels, and what she defined as a life changing experience was her trip to Nepal in the 1980s, while on leave from her teaching and before going to graduate school. It changed her life in terms of worldviews, global awareness and how unaware or self-absorbed most Americans are. She visited the jungle close to India, walking through magical forests of red and pink rhododendrons in bloom being cut for firewood to heat water for tourists. She trudged miles over the Annapurna Range, going through medieval villages to cross Thorung La Pass at just under 18,000 feet (and witness the evidence of hundreds of other trekkers with no waste management systems). She hiked into the Everest region and went up Gokoyo Peak, staying in Buddhist monasteries' and homes, sleeping above the sheep and cattle. She heard the sherpa mountain people laugh and sing at festival time, drink raksi (homemade alcohol) and saw open markets with goat heads for sale. She ate rice and dal with her hands while the locals laughed at her miserable attempts and gave her a spoon.

Cheryl marks one of her most potent memories as flying in a helicopter up to the base of Longs Peak when she was a backcountry ranger to rescue a fallen climber. Cheryl remembered, "We were dwarfed by the peaks, and one wind gust would have smashed us into the vast mountain crags in an instant. It was a humbling, but grand moment, as though I was a bird. By being in the helicopter, I could see, feel and sense the ancient power of the mountains. I realized my small place in the universe, and still feel my contribution had significance as part of the bigger whole."

Cheryl is frustrated by the sense of there not being enough time, but grateful for every day, and grateful for the opportunities to meet wonderful people throughout her life; to live in a beautiful place; and to work with fun and innovative people. The new younger employees coming into the Forest Service impress her with their knowledge, skills abilities and sense of commitment to land stewardship. She thinks that the problems they will have to contend with are greater and more complex then we have known before. But the knowledge, technology and capacity for change are greater than ever as well. She closed with these thoughts: "I wish I had great profound words at this point...there just aren't any. Some days we just have to get through, and trust things will change, get better, and there are days when it seems the universe is brilliant, colored and spinning and we are all a part of it. As I take my place as an elder, I hope for wisdom to be with those who come behind me."

Connie Millar, Senior Scientist,
Pacific Southwest Research Station

Connie's family lived in Michigan until she was three, at which time her father took a position as President of Portland State University in Portland, Oregon, where Connie (Figure 122) lived until she went away to college in Washington.

Her life has been marked by a lot of travel, both as a child and as an adult. Her family spent summers at their cottage in New Hampshire on Lake Ossipee near the White Mountains. She and her mother and older brother stayed the entire summer, while her father joined them for a few weeks each summer. From the 1950s through the 1970s, they took annual road trips from Portland to the New Hampshire cabin, visiting national parks and wilderness areas along the way.

Connie remembers a peaceful and joyous childhood, with a loving family who put family first. They spent a lot of time together, hiking, camping and skiing. Her parents were "on the same page," with family and the outdoors being their prime interests. Connie loved the outdoors and the freedom it afforded. She has early memories of running barefoot all summer

Figure 122. Connie Millar, research site on Whitewing Mountain, Sierra Nevada.

long. Those early experiences shaped her outdoor ethic that has changed for kids today. She said, "Now kids do not get that outdoor time with parents. Summer camp or maybe a week vacation is the closest."

After moving to Oregon, Connie's family bought a cabin on Forest Service land along the Metolius River, near Sisters, Oregon on the Deschutes National Forest, where they spent family time outside of summers. The tiny ranching towns such as Sisters have since developed considerably, but when Connie was young the area was rural, remote and undiscovered, and she loved discovering it. She said, "I learned by being outside."

She did not particularly like grammar or high school, and did not naturally gravitate toward math and science. Her interests were in natural history, which stemmed from her love of the outdoors and learning by observation how things worked in nature. Her father had started her hiking at the age of two. By the time she was seven or eight, she often struck out on her own, hiking all day alone, enjoying observing nature at work.

She did not have a particular affinity towards boys versus girls. During the 50s and 60s she thinks she was probably steered toward girls activities, but didn't feel it was inappropriate. She was happiest alone because usually it meant that she was in the woods where she was comfortable being, "solitary in the midst of the denizens of nature."

Her love of the outdoors and innate curiosity about how nature works first attracted her to her life profession. She did not deliberately choose a non-traditional career path, just followed her passion. She was encouraged to go into the traditional field [for women of that time] of English but chose the College of Forestry at the University of Washington in Seattle. She loves to write, and has published widely over the years, but those publications are about the research she has done, not "English essays" per se.

She said, "I have a passion for the outdoors and for understanding things, like why a particular tree is growing in a particular location. I grew up in the late 60s and early 70s, times that fostered a conservation ethic, and that further influenced my interest. I wanted to save the forests of the world. My interest was not particularly science-driven; in fact I had a fear of STEM [Science, Technology, Engineering and Math]. I was attracted to biology and natural science because I wanted to figure out how the natural world works." Her seasonal jobs during college did not have a strong science leaning, but allowed her to indulge her love of being in nature, and to share it with others.

That's not to say that she could not do math or science — in fact it was an integral part of her college studies. She first earned a Bachelor of Science in forest science at the University of Washington, in 1977, and then went to the University of California (UC), Berkeley, where she received a Master of Science in 1979. She does not remember male bias in college, and did not experience any incidents of sexual harassment during that time. She wonders if she just may not have been aware since her formative years occurred during the 1960s and 1970s, which in her experience were "very loose times."

During college she held summer jobs for four seasons as a Forest Service wilderness ranger and two seasons as a patrol leader for the Outward Bound program. In her first season as a wilderness ranger she was the only female field employee on her district. She did not feel that she was a token, and never felt discriminated against. She worked alone as a wilderness ranger, she was physically fit and felt respected.

She did witness some discrimination the following season toward another woman who worked in fire. The woman would get lost in the woods and the guys derided her. The result was that she would delay calling in that she was lost, compromising her safety.

After earning her PhD Connie worked at UC Berkeley as the project leader and research geneticist for the California Forest Germplasm Conservation Project, from 1985–1987, and has since worked for the Forest Service Research and Development branch at the Pacific Southwest Research Station, first for the Institute of Forest Genetics, then the Sierra Nevada Research Center and currently in the Ecosystem Function and Health Program. There she was a research forest ecologist and research forest geneticist through 2014, when she was promoted to her current position of senior technical scientist.

The Senior Technical (ST) designation is a unique category of Federal jobs that covers non-executive positions classified above the GS-15 level. It involves performance of highest-level research and development in the physical, biological, medical, or engineering sciences, or a closely related field.[47] Many of the Federal government's most renowned scientists and engineers serve in ST positions. It is the highest designation bestowed on a researcher employed in federal service. Only 470 ST positions are allocated across the federal agencies, twelve of them in the Forest Service at the time of 2016. Connie was at that time the first and so far only female Forest Service scientist to have been appointed to the ST level.

A paucity of guidance when she started her Forest Service career with her first research job afforded her a lot of freedom to define her work and do what she wanted. She barged ahead and did what she thought was important. Her interests in various areas of science varied over time. Her interests gradually turned from genetics to studying other aspects of plants, mammals and geomorphology. She said, "I had the freedom to reinvent myself as a different sort of scientist as my interests changed. Now it seems like a miracle that working for the Forest Service has fit so well. I have been able to sculpt my career in a way that works for me and for those I serve in the National Forest System (NFS)."

By that she meant that while Forest Service Research is a separate branch with its own research goals, the research done by its employees also serves to inform the NFS branch. Connie's interests lie between doing basic research in natural history science and the other side of serving national forest needs. To that end, early in her career, she spent a three months sabbatical working on a ranger district to gain a first-hand understanding of needs at that level. She and her research colleagues went into the field with ranger district employees to understand how scientists could support their needs. During the early 2000s, she and her colleagues took time out from field research and spent a year talking to the National Leadership Team and others about how to take a resource approach to climate change. That effort led to development of many climate-change and climate-adaptation programs the agency now has, including a climate scorecard and a national strategic plan.

Mid-career, before remote workstations were common, Connie decided to take her work to the mountains. That practice has grown into a field season of about five months each year that she and her team spend conducting research on the east side of the Sierra Nevada, based in the Mono Basin of eastern California and roaming farther east into the mountains of the Great Basin (figures 123 and 124).

Unlike the NFS, where mobility is encouraged in order to promote, research branch scientists usually are not mobile, but are able to stay and promote in place. Connie and her lab technician and statistician have been together since the 1970s. She said, "Their strengths complement the pieces I don't do well. I am not particularly detail-oriented (except in things I really like) or skilled in quantitative analysis while they are. We each have qualities that contribute to productive collaborations."

Ironically, the freedom to develop her program as she saw fit left her feeling inadequate at times. She felt that she was not a "real" scientist because she wanted to climb and hike versus being in a lab. Mid-career she had a more gender-based sense

Figure 123. Connie at Lundy Canyon ridge top looking east. Mono Lake (and her home) in the basin below.

Figure 124. Connie resting on the ridge top of Lundy Canyon after long field day.

that woman scientists talk differently than male scientists and she felt like she had to act and talk like a (male) scientist. Some female scientists were not soft and behaved more like men, not emotive or intuitive, which did not help Connie's self-image as a scientist.

In 1978, psychologists Pauline Rose Clance and Suzanne Imes, described the Imposter Phenomenon, 48 [or Imposter Syndrome] "To designate an internal experience of intellectual phonies, which appears to be particularly prevalent and intense among a select sample of high achieving women." Women afflicted with this syndrome, especially successful women, feel that their success is attributed to someone else — that their accomplishments are not earned, but that some deception or hoax brings about promotions or praise. They are fearful that they are not responsible for their own success.

Connie learned about this syndrome during a 2016 conference that examined the reasons why there are still relatively few women in research. The concept resonated with her. She has felt the same many times throughout her career, which illustrates the point, considering the magnitude of her lifetime accomplishments. She later accepted herself and her differences, and is comfortable with being accessible to other women like her.

There is no typical day of work in Connie's world. Her work has varied widely over the course of her career, and it varies by season and day. During the field season (June through October), she starts the day at 7:00 a.m. with office work, and then prepares for the field day. She usually leaves between 8:00 and 9:00 and hikes five to ten miles daily. She returns by 6:00 p.m. and usually does another hour of office work.

Fieldwork, based from her house and the ranger station office in the Mono Basin, includes tree-ring coring (figures 125 and 126), monitoring small mammals such as Belding ground squirrel, American pika and marmot (figures 127 and 128) and analyzing bark beetle, mortality and forest growth within forest plots, assessing alpine vegetation and thermistor (small temperature recorders) deployment and exchange (Figure 129).

Figure 125. Connie coring a dead bristlecone pine in the North Snake Range, Nevada.

Figure 126. Connie conducts tree-ring studies on ancient limber pines in Sweetwater Mountains.

Figure 127. Connie surveying for American pika on Bald Mountain, Shoshone Range, Nevada.

Figure 128. Connie recording info on locations of American pika and conifer occurrences.

Figure 129. Connie exchanging thermistors as part of forest ecology studies.

Thermistors are used in many areas of Connie's research: to determine micro-climates of pika habitat (interiors of talus fields where they live), and which are air-conditioned by unusual circulation patterns, to study how rock glaciers develop permafrost at elevations up to 1,000 meters lower than general permafrost levels (Figure 130), to study thermal relations of subalpine forest trees and their climate-change induced range shifts, to study tree line thermal limits and, most recently, how "krummholz" (shrubby) forms of pines are able to transcend those limits and by how much. They are used to study microclimates of alpine plants, to study how temperatures are (if they are) changing in alpine environments and how that relates to alpine vegetation changes. Her work takes place in the subalpine to alpine zones, usually at and above tree line (Figure 131). Her geographic focus is the mountainous Great Basin, from the eastern Sierra Nevada through the mountains of Nevada, southern Oregon and western Utah (Figure 132). In winter (November through May) she sits in a tiny cubicle in

Figure 130. Connie calling in a data report from Conness rock glacier after exchanging thermistors.

Albany "chained to my computer from 7:30 a.m. to 6:00 p.m., with an hour and a half break for exercise." During those times she likes to think she is analyzing her data and writing her research publications, but many interruptions intervene.

Though Connie has been fortunate not to have experienced any blatant sex discrimination throughout her career, she has faced what she calls "institutional challenges," meaning that she sometimes gets frustrated by the inevitable bureaucracy that comes with working for a large institution. Budgetary restrictions that curtail travel and essential job-related training and workshops can hamper work progress. Mandatory-training sessions, important but not directly related to her work (safety, computer security, sexual harassment, defensive driving) come in droves thanks to the ease of delivering them electronically. They take up time that could be spent in the field. At the same time, Connie is proud of the agency for developing discrimination training that has

Figure 131. Connie on Toiyabe Crest studying bighorn sheep habitat and behavior.

Figure 132. Connie on Parker Bench, eastern Sierra Nevada, surveying rock glaciers.

benefitted women and all employees, that teaches that each person is a unique individual with qualities that contribute to an enriched whole. That goes hand-in-hand with the "family" culture fostered within the agency. She has stayed with the Forest Service because she loves the land and the land stewardship that the agency practices. She loves the fulfilling career it has offered her, especially the freedom to invent her self and to make her research work relevant to work on the ground. She said, "I only have one life, and I identify with my career, in a good way. I would be a different person if I worked in a lab. If I had it to do over, I would do it again. The flexibility has allowed me to grow."

Now Connie is driven by having the choice of what she does with her remaining years on this job. She views life stages as occurring in "trimesters," and places herself currently in the late trimester beginning at menopause and moving into another part of life, retirement. She wants to do something related but different, focusing on her own research and field work, writing about it and giving herself permission to make choices and to say no to requests that do not fit her mission. Another "big passion" is to give back by serving in an advisory capacity, shouldering the responsibility to share her wisdom. Her greatest goals are, "To do no harm, take the high road, respect all life, remember etiquette, serve, choose battles wisely and give thanks."

Reflecting on her experiences over the course of her career, Connie said, "I can't think of any set of experiences that is more extraordinary than the glory of being in the high mountains on a daily basis, unraveling the detective stories of how plants and animals got to be where they are, the miracle of evolution that produced organic life on Earth, the great blessing that I am fit and healthy enough to continue to bask in this world and the great fortune to have a job that allows and encourages this."

Roberta Page, Law Enforcement Officer

Roberta (Figure 133) has always lived and worked in Missouri. Her Forest Service career took her to different locations, and then she moved back to her hometown area in the country at Mountain Grove, Missouri.

She said she was a major tomboy as a child, bored by dressing up and playing with dolls. She roamed her family's farm with her two older brothers playing sports and shooting BB guns. The family also spent time at a lake, skiing and camping. She enjoyed doing construction with her dad, who was always remodeling and improving their home, and she learned early how to use a hammer. Her sister, Karen, had muscular dystrophy so Roberta also spent a lot of time being Karen's best playmate and friend. There was a strong church influence in Roberta's life, as her mother was a Sunday school teacher.

Roberta loved horses, dogs and reading. Her favorite subjects in school were history, literature, art, mythology and math. She was good at math, and was also very interested in science, though she felt she was "not the best at it." She was

Figure 133. Roberta Page with canine Breaton.

a minority in some of her high school classes. She hated the mandatory home economics class, especially the sewing part. She said, "I despised all the little house work skills they taught, with the exception of cooking. I absolutely love to cook. I believe it is because it is somewhat math and science combined."

Unlike most of the other law enforcement officers (LEOs) Roberta knew, she did not have a college degree. She graduated from Mountain Grove High School in 1976, and was married that fall to her husband, Rick Page.

She was introduced to outdoor work with the Forest Service in 1979, when she worked on a YACC crew on the Houston Ranger District of the Mark Twain National Forest. Although she only worked a few months, she loved working in the outdoors. She worked at various jobs including the YACC, construction and factories (where she worked her way up to supervisor positions). In 1987, she was hired as a correctional officer for the State of Missouri, where she worked until 1990. She was then hired as correctional officer for the Department of Justice, Federal Bureau of Prisons. She worked as a correctional officer then a senior correctional officer until late 1997. While working for the Department of Justice, a coworker asked her where she would rather work. She told him she would rather work for the Forest Service, which she eventually did. She first worked as security for the Social Security Office of Hearing and Appeals and the U.S. Attorney's Office until the Forest Service hired her. She became a law enforcement officer for the Forest Service in July 1998, and worked for them until January 31, 2014.

Now retired, Roberta said that what she loved most about her job was working with her K-9 partners (Figure 134). She loved working narcotics with them, going to schools and doing demonstrations for the kids and hiking through the woods with the dogs.

Roberta loved the variety that being a LEO provided. She would get ready for her shift, walk out the door to her vehicle and decide "east or west today." That was if she had no complaints to investigate instead of patrolling. During the summer,

Figure 134. Roberta and Tuomo after NAPWDA Certification, a nationally recognized narcotic certification.

she spent most of her time at the river and camping areas. In fall she worked hunting camps and dealt with off-highway vehicle (OHV) problems, fall fire season and arson investigations. In winter firewood theft was her main concern. Spring began another fire season, hunting season and more OHV problems.

Roberta had worked a non-traditional job for many years prior to being employed by the Forest Service, and said, "Working in the male-dominated field was nothing new to me." At first she felt that she did not belong for a different reason, because of her stronger previous law enforcement employment. She said, "The Department of Justice had very direct and strong leadership. The Forest Service was very wishy-washy on their regulations — they used them somewhat like guidelines. When I went to the academy I was told to give people the benefit of the doubt. It was quite a learning curve for me. I was written up for being assertive in law enforcement. I was told that if someone did not want to comply with a regulation and was getting belligerent, I should just walk away."

The most challenging thing she had to face came later in her career when the "New Face of the LE&I" reversed the techniques Roberta had learned in the academy and had practiced for years. Roberta said, "After LE&I decided to become 'real law enforcement' I was chastised for not being aggressive enough. So I did what they asked. The new regime in D.C. wanted tickets written for every little transgression. As a female, if I had to become assertive, as you have to do in some situations, it was considered rude. I see the challenges of the Forest Service female LE officers as trying to be good officers without being held to the double standard. The supervisors need to be aware that in some areas men do not like females in authority, and they will automatically take offense if you tell them they are violating rules."

As she drew closer to retirement, even though she remained physically fit, her last captain was condescending to her. She said, "I don't know if it was my age, me being a female, or the fact that I had more knowledge of the job than he did." She had always been included in special operations, until he showed up. He assigned male officers to the special ops. All of her other captains had gone to work with her on occasion; this one did not. His office in Rolla was not far from hers, but when Roberta retired in 2014, he had to call her to get directions to her office. She said, "So I guess the special challenge I faced was trying to find a reason to stay until my maximum age — I couldn't." She said:

> I am afraid the LE&I program is headed in the wrong direction and I am glad that I was able to retire. When I started,

I was instructed to give people the chance to "do the right thing" before writing violation notices, and to use the "reasonable thinking person" as a guideline. When I left, there were quotas for violation notices (although they will deny it) and you were judged on how many violation notices you wrote. Nobody cared if you helped someone in need, if you found a lost person, if you put on a smile and made someone's day better. They wanted to know if you wrote the lady with six kids in the beat up truck a $75.00 fine for not paying a two-dollar day use fee. If not why? Or why didn't you write the person a violation notice for driving on a user-created road that actually looked better than the Forest Service road next to it? It is all about statistics and nothing about the human element.

Still, Roberta has good memories of her Forest Service career. Working in New Orleans in the aftermath of Katrina was her top extraordinary work experience. Working undercover at the 1999 Rainbow Gathering was number two (It was her first gathering and she had "*never* seen anything like it"). She counted representing the Region 9 K-9 program in Washington, D.C. as the third. She had never been to D.C. and she was amazed at the beauty Figure 135).

Figure 135. Roberta's Forest Service K9 Ruutu in Washington, D.C., 2010.

Working in details in some of the most beautiful places in the United States was also high on her list. She said, "That is something I liked about the Forest Service. I was able to work for two to four weeks in different states. If I liked the area, my family and I would vacation there later."

Roberta has kept her hand in law enforcement. She still trains K-9s and is getting her Veteran Police Officer License for the State of Missouri. She has had job offers from the local police department and a couple of counties that she worked closely with as a Forest Service LEO. She holds law enforcement commissions in three counties as a Reserve K9 Officer and gets called out for special operations where a K9 is needed.

She lives with her husband of over forty years on fourteen acres, where they built a new house, and where she has a dog kennel with her retired K-9, his son, which is her new K-9 and four other dogs from the working line. She and her husband belong to a Law Enforcement Motorcycle Club, and they raise money for different charities.

Saddened by the circumstances that led her to retire early, Roberta said, "I really and truly loved my job, and was proud to say whom I worked for."

Iral Ragenovich, Regional Entomologist, Forest Health Protection, State & Private Forestry

Lewistown, Montana, an agricultural community of about 6,000 in central Montana, was Iral Ragenovich's childhood home. Iral (Figure 136) described the town as a pretty spot; seven small mountain ranges form a basin, with Lewistown sitting in that basin.

Iral characterized herself as a classic tomboy. Like most girls of the time, she did play with dolls when she was young, but most often girls' activities and especially their conversations bored her. She and her sister made up stories involving their dolls that had a creative twist. Iral said, "My sister and I had dolls that were run away princesses that sailed away in a muffin tin." Their imagination extended to other games

Figure 136. Iral Ragenovich.

as well. She said, "My sister and I and my friend, Claudia, played a game called 'Silver and Gold.' Claudia was a horse called Silver, my sister was a horse called Gold, and I was the girl who owned them; we went on wonderful adventures together."

Iral preferred being outside and active, often playing with her two brothers. Sometimes they played softball but she preferred individual activities to team sports. Most often, she and her brothers rode their bikes all over town and out into the country.

In high school, college and even later, Iral was not interested in the topics that dominated most women's conversations. She said:

> Conversations often centered on clothes, and boys, and planning for marriage, and later husbands and kids. Not that I mind clothes — I have always had a reputation of being a 'clothes horse,' I just don't spend a lot of time talking about them. I don't like conversations that approach gossip, and kids hold no interest to me, whatsoever, because I cannot

even begin to relate. Those are often the only subjects many women have, probably because those are central to their life and the most familiar. Guys seemed to talk more about other things – granted, sports – but also about things related to sports, sometimes people, often ideas and events, and how to figure out how to make things work. Boys were also more engaged in activities, rather than sitting and talking and I wanted to be doing things, not sitting and talking.

Archery was her dominant sport during junior high and high school. She was in the archery club in town and participated in both indoor and outdoor archery competitions throughout the state. She said, "I particularly like the outdoor courses that follow a trail through the woods and create a varied terrain; sometimes you shoot uphill, sometimes across a gully and sometimes between some trees. I don't remember anymore how many trophies I had; they are packed away somewhere."

Iral and her siblings all had chores. The girls helped clean house and wash dishes. Mowing grass and shoveling snow usually fell to her brothers, but she would snag the opportunity to do those whenever she could.

The property out in the country that they referred to as "the ranch" was often a Sunday destination. It was an 80-acre remnant of the original ranch that her grandparents kept when they sold the rest. Iral reminisced, "We would pack up some food and drive out to spend the day hiking in the trees and hills, climbing Alaska Bench or Shook Bench, or target practice (with my dad supervising – he taught us to use and respect guns). My dad would put the watermelon in the spring so it would get cold."

There was very little formal or organized recreation such as camping or skiing. Her dad, a civil engineer for the Montana State Highway Department and also a private land surveyor, pretty much worked all the time and did not take time off for vacations. Iral said, "The closest in that regard, was every summer after school was out, mom and us four kids would go to Kansas for about six weeks to visit mom's folks and family."

Commitment to family, to God and respecting elders and others were strong family values. Her parents placed a high value on education and taught their children to do their best in whatever they did, to have a strong work ethic and to be self-reliant.

School was something you did because you had to, and Iral could not think of any school classes that she particularly enjoyed, other than physical education. She thought reading was okay, and in high school she liked biology. She was ambivalent about other classes. The ones she disliked the most throughout her school years were math classes. She said, "I don't remember ever liking math — it was always difficult. Converting the real world into numbers or numbers into the real world did not equate, which is kind of an oddity coming from a family with a strong inclination for math. When I was growing up, my dad and sister and brothers would actually sit at the kitchen table after supper and work math problems (with a slide rule no less) for the fun of it! Not me." Her family has a strong representation in the civil and mechanical engineering fields.

She had a stubborn streak in some respects, which she demonstrated when it came to choosing her career. She said, "To be honest, I specifically looked at professions that were non-traditional for women. At the time, women were expected to go into the secretarial, teaching or nursing fields. I wanted to do something that I was told I 'could not do because I was a woman' just to show them that I could." In addition, most of the expected careers for women involved working with people, and as a self-proclaimed introvert, Iral was not comfortable with that. For a while she thought she might want to be a veterinarian, but she was only an average student and did not want to be in college for as long as it would take to be a veterinarian. She had read books by people who had spent summers in lookout towers, and the idea of being alone most of the time on the top of a mountain really appealed to her. She said, "From that came the idea of forestry, even though, at the time, I had only a limited concept of what that involved — lookout towers and fighting forest fires — but it was a field that no other girls were considering, and one that challenged the traditional idea of a woman's career." She was not interested in many other professions, although she did consider biological illustration.

She faced a challenge right away, when her father objected to her career choice. He was adamant that she should not study forestry and they used to have some serious arguments about it. It was not because he was strictly against her being in a male oriented field (her sister was taking civil engineering and he was okay with that), but that he was concerned about what his daughter would be exposed to, and that she would not be able to find a job. She acknowledged that given the times, he might have had legitimate concerns. She counts her dad deciding it was okay for her to take forestry as one of her top extraordinary life experiences, and described how he came to terms with her choice:

> My father's only concept of forestry was fighting forest fires. To his credit he talked to someone who was a forester (it probably came up in a conversation — I can imagine him saying, "My daughter has this crazy idea about wanting to be a forester...") and that forester actually said it was a good idea, that some of the best tree planters he had were women because they took the time to make sure the trees were planted correctly rather than just jamming them in the ground. So now forestry included fire fighting and tree planting and that seemed to be more palatable. When I came home from college (my undergraduate degree was in biology) during my senior year, not only had my dad changed his mind, but you would have thought it was his idea. After we graduated, he would often introduce my sister and I, as "my daughter the engineer" and "my daughter the forester." He got a kick out of that, because we had accomplished something unusual.

After graduating with a biology degree from Walla Walla College in 1970, Iral was accepted into the Masters of Forestry program at the Stephan F. Austin State University (SFA) in Texas. While working on her graduate degree, she also completed the course work to obtain a second undergraduate degree in forestry. She received her graduate degree in forestry in

1973. As a woman in the early seventies, she was a minority in her classes. She said, "Even as an undergraduate, when I was a biology major, women in biology were also a minority, although not to the extreme that they were in forestry school. In many of my biology classes I was one of three or four women. I was often the only one or one of two, in my forestry classes. I noticed that there were very few, or that I was the only one, but I never felt like I did not belong there. I don't know exactly how to describe it; it just was what it was. I suppose, if anything, I stood out and generated more attention than if I was just another of the 'crowd'. Certainly, even now, people remember me." She was the third woman to get a bachelor's in forestry and the first woman to get a master's in forestry from SFA. Her master's thesis was on preventive and remedial control of southern pine beetle (SPB) and ips bark beetles using a variety of insecticides. It was through working on her thesis on southern pine beetle that she became involved in forest entomology. She said, "I often say I got into forest entomology through the backdoor. When I first started in graduate school, I planned to go into silviculture. After my first year, my major professor left for another university, and the forest entomology professor was willing to take me on; hence, the switch to forest entomology. After I graduated I applied to both the forestry and entomology rosters through the Office of Personnel Management. I was offered a position as a forest entomologist with the Forest Service in Pineville, Louisiana. I accepted, and have never regretted it." Indeed, Iral's Forest Service career has spanned over forty years, so far.

Iral first worked for the Forest Service as a seasonal during summer breaks from the university. In 1971, she was one of two of the first women ever hired to work on timber inventory crews in Region 1 on the Lewis and Clark National Forest in Montana. The following summer she worked as a timber inventory crew leader in Region 1 on the St. Joe National Forest in Idaho.

During the very cold Montana fall and winter immediately following her graduation, she helped her dad with land surveying, then in the spring of 1974, she worked at Texas A&M as a lab technician working on a southern pine beetle project testing male beetle response to pheromones. In May of that year, she began her permanent career with the Forest Service, as the first female forest entomologist hired by the State and Private Forestry, Forest Pest Control Branch (now Forest Health Protection). The position was located in the Pineville, Louisiana field office for the Southeastern Area. She said that at the time the Forest Service was going through change. Hiring women into the forester and field positions was new to the agency. Whether I was hired because they had to show they were hiring women, or if I was hired as a "trial balloon" to see if it would work, or both, I do know that my gender was a consideration, both when I was hired as one of two women on the inventory crews, and certainly when I was selected as a forest

Figure 137. Iral working in Mobile-Tensaw River Basin, Alabama.

entomologist. Following the interview she had before she was hired permanently, she said, "I felt I was being hired because I was a woman and almost turned the job down because of that. I wanted to be hired because I was a forest entomologist, not because of my gender. My major professor's advice was that even if that were the case, I should 'take the job and show them you are the best damn entomologist they have,' and I think I have done that. I know I am well respected in my field." She said her major professor deserves recognition for having had a significant impact in her life, and she said, "Early on he was the one person who believed in me and encouraged me in graduate school. He was truly a mentor. Following his advice, I never looked at being a token as a negative, as much as it was an opportunity to show what I could do, and make myself visible. And I know that how I performed has paved an easier road for many women coming after me."

While in her first position she worked primarily on aerial pest detection surveys, evaluating and providing advice on managing forests for southern pine beetle and a forest tent caterpillar spray project in the Mobile-Tensaw River Basin in Alabama (Figure 137).

Working on the forest tent caterpillar spray project in the swamp in Alabama was another of her most extraordinary experiences. She said, "It was the first major project that I worked on so maybe it is because the experience was so fresh in my impressionable mind. But the whole ecosystem was unusual and fascinating as well. And the people I worked with were interesting and fun." The Pineville Field Office was responsible for insect and disease monitoring and management of the six Gulf Coast states of Texas, Oklahoma, Arkansas, Louisiana, Mississippi and Alabama.

In 1976, Iral moved to the Asheville Field Office in Asheville, North Carolina. Her duties there were similar, but she worked on slightly different insects. In addition to southern pine beetle, she worked on insects such as the balsam wooley adelgid and fall cankerworm, and also helped with seed and cone insects in southern seed orchards (Figure 138).

Figure 138. Iral — balsam wooley adelgid plots, Mount Mitchell, North Carolina.

The Asheville Field Office served the states of Virginia, North and South Carolina, Kentucky, Tennessee, Georgia and Florida. This covered a broad variety of ecosystems, ranging from the coastal plains to the Piedmont to the Appalachian Mountains.

Two years later she transferred to Region 3, the Southwestern Regional Office in Albuquerque, New Mexico. There she did aerial detection surveys, worked on western spruce budworm and Pandora moth spray projects and worked with other western insects such as the mountain pine beetle, ips and Douglas-fir beetle. During her time in Region 3, Iral also held other positions. She was the pesticide coordinator providing advice on pesticide management and use and providing training for pesticide applicator certification for Forest Service personnel. As the regional timber inventory specialist, she prepared timber inventory contracts and helped the forest silviculturist develop the inventory program for the coming summer.

In 1984, she transferred to the Pacific Northwest Regional Office as an entomologist, where she again worked on aerial detection surveys, ran a western spruce budworm spray project, conducted biological evaluations and provided training to all federal land managers (Figure 139).

She also worked on insect management related environmental analyses and impact statements. Three years later she became the supervisory entomologist for the region and remained in that position, supervising six to seven entomologists and technicians, until the agency created "zone" or field offices and moved many of the entomologists to field locations. Iral remained in her current position as the regional entomologist. In that position, although she sometimes assists the field offices with biological evaluations and with training, most of her responsibilities are now managing and implementing regional

Figure 139. Iral — Wenatchee, Washington, sirex wood wasp project.

programs and projects, and managing the forest health related grants and agreements with the states and universities. She has also worked on a number of special projects, some national or even international in scope, including work with the Sukachev Institute of Forestry in Krasnoyarsk Russia in developing pheromone for Siberian moth.

Iral said, "What I have liked most about my various positions has been the diversity of insect problems, the opportunity to work in a variety of ecosystems, and the travel. I have had the opportunity to be involved in projects at the local, regional and national level." Forest Service entomologists provide service to the NFS, but also provide technical assistance and training to forest and land managers for all federal land managers, such as the National Park Service, Bureau of land Management, and the Department of Defense. They also work very closely with states' departments of forestry or natural resources that are the primary contacts for state and private land managers. This provides exposure to not only a variety of insect problems, but also a variety of land management objectives, some of which are very different from the NFS.

Though she got a job, and a good one, Iral did face some challenges. She said that probably all women in non-traditional fields shared many of the challenges she faced, particularly during those times, in the early seventies. Early on, she felt she had to continually prove herself. There was an attitude, or belief, that she would not be able to hold up her end of the workload, or could not do it because she was a woman. The men assumed they would have to do extra work because she would not be able to pull her own weight. Iral never noticed it, but some did confess it to her later on, "After they had worked with me and gotten accustomed to working with me." One of the men she worked with would never tell his wife when he was traveling with her because it apparently caused problems for him at home if his wife knew he was traveling with a woman. Iral said that while she did not feel excluded during times on the job, she may have been excluded from some of the after work activities. But, she said, "As I mentioned, I tend to be a loner and I don't know that I would have enjoyed them anyway, and if I were excluded, I never knew about it"

A woman in a Forest Service uniform was unusual. When working alone, if she stopped at a restaurant for lunch, she would sit with her uniform patch to the wall, or folks would openly stare at her; she hoped it was because she fascinated them. Iral told a story that vividly illustrates the attitude of that time, particularly in the Deep South:

> One time I was visiting one of the Ranger Districts in Louisiana, looking at and evaluating some of the southern pine beetle problems they had going on. We stopped at a little country store for something cold to drink — Louisiana being rather hot and humid in the summer. It was just a small place — sort of long and narrow — with a single counter running the length and some canned goods and notions along the back wall behind the counter. A refrigerated chest-type cooler near the front of the store contained the cold drinks. The man behind the counter was small and older with thick glasses and a shock of white hair. It was dim and dark, especially coming in from the bright sunlight; a single bulb hanging from a wire was the light. The two foresters I was with paid for their drinks and waited for me by the door. I tried to hand the proprietor my money but he just stood there completely immobile and unresponsive. I was afraid he had gone into an epileptic seizure, similar to the kind a girl at school used to have. I felt a bit of a panic and turned to look at the two guys I was with, but they were talking and did not seem to notice, so I turned back to the shopkeeper, thinking if I said something I might find out if he could respond. So I held out my hand and said, "here is my quarter" (I actually think the soda did cost only a quarter back then). His response was, "What's the Forest Service coming to? First they hire niggers, and now they are hiring women." I was so relieved that he was all right that the impact and implication of what he had said did not even sink in until quite a bit later.

Other stories that stand out revolved around being a white female working with a black guy, and trying to understand public perceptions in the South in that regard. Some of her more entertaining challenges had to do not with people, but with things like the logistics of how to relieve one's self in the middle of a swamp (at flood stage the water was 12–15 feet deep so all of the work was done entirely out of aluminum boats or canoes — even the water moccasins had trouble finding a high, dry place to hang out).

One misconception of that time was that any woman in forestry, or other male-dominated field, must have been trying to find a husband (or worse, be "easy"). After she graduated the comments were: "Now that you have your MF, you can work on your Mrs." Another comment was, "If you are planning to get married, don't bother to apply for the job," because they did not want to invest time in someone who was going to get married and then quit. Of course that same sentiment did not extend to men.

Iral said that her biggest challenge was probably a supervisor who did everything he could to keep her from succeeding. She said, "This was probably the darkest period of my career and it forced me to stiffen my resolve and really believe in myself. I was promoted into the position of supervisory entomologist, the position my immediate supervisor had held prior to his promotion to the director of the unit. It was not entirely because I was a woman — everyone who worked for him disliked him. Some of the other folks on the staff held an intense animosity for him. In my case, I believe it was that he did not want someone else, and especially a woman, to show that they could succeed in his former position equally as well, or even better than he himself had done." She said that another female entomologist who worked for the same supervisor left the Forest Service, partly because of the way she had been treated. Iral said she did not think that it was entirely because her colleague was a woman, as much as it was the way the supervisor used his position to control and manipulate the people who worked for him. He treated some of the guys similarly, but they could not afford to quit because of family and other obligations, or they simply moved, if they had that option. The bright spot and where the agency benefited from that situation was that it gained a number of folks who were very good supervisors, because they learned, through trial by fire and experience, what it was to be a bad supervisor. The dark blot on the agency was that it was a prime example of how the agency ignored or avoided dealing with a bad personnel situation. She said, "I stayed with the agency because I believed in the agency, and could separate my career and the mission of the agency from the actions of such individuals. I made the decision long ago not to relinquish the control of my life and career to individuals. If I spent time dwelling on how I was treated, then I would, by default, be letting them control and decide my life — so I looked past them and moved on."

All in all, Iral says her passion is still her job. She said, "I like to think — and I certainly hope — that the challenges that I faced early on paved the way for acceptance for others that followed."

Nearing the end of her career, her greatest goal related to work is to complete a revision of the Western Forest Insects book. It is a job she took on several years ago, in the role of the technical editor with the help of a number of others working on it with her. It turned out to be a major undertaking that will take a lot of effort to complete, and is one thing Iral wants to complete and leave as a legacy for future forest entomologists.

Her sister is the closest person to her, and her closest friend. That makes a personal goal of hers of visiting all fifty-nine major national parks together particularly worthwhile. Her family — her sister and brothers, and when they were alive, her parents, especially her mother, are the most important people in her life. She said her colleagues at work are an important factor, just because of how much time she spends with them, but she seldom associates with them outside of work. She offered the following insight into her personality:

> I tend to keep to myself. I fit more on the introvert scale and have only a few close friends. I am comfortable in my own company and do not seek to socialize. I never married although I almost came close once, but I could not make a commitment that down the road might force me to make a decision about my career. Marriage was not an important

goal in my life and my career has always been first. Kids were never on the radar — the "mother" concept has always been foreign to me. I think I was behind the door when they were passing out that particular instinct. I know that my situation is probably unique in many ways because I drew a conscious line for my career that excluded options in my personal life. Part of this is my preference for my career — but I am sure it is also a function of the thinking of the time — whether as a result of the way I was raised, or the response and actions of others. Many other women who have a different vision of their life have had to face different challenges of career and family, and how to blend and balance those.

Asked whether she felt that working for the Forest Service has changed her, she said, "It would, because we are who we are through our experiences. Over forty-two years I have certainly changed — become surer of myself, have self-reliance and self-respect and try to make decisions that incorporate all views — nothing is cut and dried. But is that because of the Forest Service specifically? Some of those changes might have occurred just in the process of growing and maturing over time, regardless of what I was doing. I do think it gave me opportunities to travel and work in different parts of the country, experience different cultures and ecosystems and meet and learn from many interesting people. And it has helped me to see that many factors are at play in almost everything we do. Would I have had some of those same experiences and opportunities if my career had taken a different path? Perhaps a few, but I think not likely to the same extent as I have had working for the Forest Service"

∼

Tina Terrell, National Director of the Job Corps

Born and raised in the city of Philadelphia, Pennsylvania, Tina (Figure 140) calls herself an "East Coast Yankee." She loves the east coast and the "City of Brotherly Love."

Tina went to Catholic school from second grade through high school. Her family was not Catholic, but the school provided the best education in the Philadelphia area. She completed

Figure 140. Tina Terrell.

all the high school requirements for graduation in 1982, and was in the top ten percent of her class, but did not receive a diploma at that time, as her family could not pay all the tuition bills from high school. Her high school transcripts were not released to the college she wanted to attend. Tina went to work since she was not able to attend college, to take care of her family and to help her sister who was a year older than her to stay in college. From September 1982 to May 1983, she made enough money to pay the $2,500 high school debt and get out of high school, to help her family pay the bills and to send funds to her sister in college. For about three to four months straight, she worked every day. She was responsible for working sixty hours every two weeks, but decided she needed to work more hours. She averaged 100–120 hours for a period of time, working nearly double hours at her McDonalds job. This life lesson taught Tina the "biggest aspect of being: determination to succeed and a strong work ethic."

Tina received her high school diploma and had her high school transcripts sent to the only college she wanted to attend — Penn State, which she attended from 1983–1987. She said, "I knew from the age of two it was the only college for me, and I applied to no other colleges or universities." She initially studied environmental resources management but changed her major to forestry in 1984. Her family was not able to support her, but through work, loans and grants, she made it through college. After she changed her major to forestry, she worked at a Forest Service job in the summer and another job for twenty hours per week during school. She graduated in 1987, and paid off her student loans by 1993. She is currently the only African American, male or female, to graduate from Penn State with a bachelor's degree in forestry.

As a child, Tina loved the outdoors — her mom had to pull her indoors all the time. She shares a love of sports with her mother. She understands how to play the "game," whatever the game.

In high school she liked the arts (drawing), science and American history. She enjoyed learning about how the American society got established. In college she loved science, especially organic chemistry, and economics, business and communications.

Tina's family is close-knit. She is the middle of two sisters, and has lots of aunts and uncles; her mother had twenty-one brothers and sisters.

The family lived in the urban city, a half block from the projects, with lots of concrete and asphalt, where it was not safe after dark. Tina learned to be defensive and direct with people. You would never know now that she was once extremely shy.

Education was of utmost importance to her family. Her mother worked hard as a salesperson in a department store for thirty-seven years to ensure her daughters were well educated.

Her mother was not a manager, but many managers would speak to her mother regarding how to sell, managing the cash register, or overseeing the ordering of goods.

Tina's mother provided such a high level of customer service to the public and to her supervisors in her job that customers would prefer waiting in line to receive her service over others'. Her mother's line was always longer than other sales personnel on the counter. Managers asked her mother for advice. She epitomized customer service in life; Tina learned that value from her mother, as well as self-reliance and confidence.

Tina fell in love with nature through her mother, who loves viewing nature. Her mother does not like to be out in nature as it is too hot for her — she likes to view it from the car. Tina did not go camping until she worked for the Forest Service. She was attracted to her profession because she loved the outdoors and going out to see what the Forest Service does. Now helping others is what Tina loves most about her job. She feels she learned that from her mother's example of providing customer service.

She loves solving problems, especially dealing with youth and helping them find their way in life, as she feels she was helped as a youth. She said, "Our job in moving society is to help young people find their way in life; open doors for them, don't close them. People can do anything they put their mind to."

After her first year at Penn State, Tina's mentor shared a contact at the Forest Service Northeastern Research Station. The research station was seeking minorities in forestry to work for the summer. Tina got a summer job doing forestry data manipulation work in the office. She had no car, and commuted one hour each way by bus and train.

Another mentor and supervisor sent Tina out to learn about forestry in the field. She went to Massachusetts for two weeks and inventoried trees. She had to get gear for the field for the first time ever, which included boots, a checkered shirt, thick socks and a hard hat. She was not familiar with the type of boots to buy for working in the forest and got construction boots instead of boots with Vibram soles that are typically used for forestry work, and the soles were not as stable as they should have been.

She counted trees every day and loved it. She lived in a hotel for two weeks and worked in an area that had three colleges/universities near the Boston area. She learned to conduct utilization studies and valuation of trees. A contract logger was cutting trees on one project, and when Tina saw this activity she thought she wanted to be a tree faller. Inspired by this experience, that is when she decided to change her major to forestry. Her mother and her mentor supported her decision. Her mother's first words when she talked to her that evening after seeing a tree fall were, "whatever you want to do." In 1984, at the end of the summer her mentor offered her a co-op position, which meant continuing to work every summer at the Forest Service research station and upon graduation (assuming a satisfactory performance rating), she would be offered a permanent position. She graduated in December 1987, and worked at the research station until 1989.

In 1989, Tina applied for jobs in the west, all in California, and received eight offers. Her final decision was between the Sequoia and Six Rivers National Forests. She chose the Six Rivers and became their small sales officer. It was her first time west of the Mississippi River. She lived in Crescent City, California in the most northern part of the State in a town of 2500 with only four black people, including her. She loved it. She located and planned timber sales, both small and mid-sized. She was also responsible for commercial and personal botanical products like mushrooms or bear grass.

It was on the Six Rivers in 1990, that Tina met one of her best mentors ever, her supervisor, Ken Wills, who was a "good old boy" who took her under his wing and "made sure that I thrived, not just survived." At that time another of her mentors recommended that she take on collateral duties as the African-American Program Manager, which she did. This experience opened Tina's eyes to outreach, recruitment and her eventual passion — helping the Forest Service represent the public we serve. She began to want more challenges and her mentors saw leadership potential in her and so she took some leadership courses. In 1993, she attended the Blacks in Government (BIG) Conference in Los Angeles to help recruit minorities into positions with the Forest Service. It was one of many conferences she would attend from then on. She usually attends at least two per year, such as the Society of American Foresters-SAF, Minorities in Agriculture, Natural

Resources or Related Sciences-MANRRS, BIG, National Association for the Advancement of Colored People-NAACP and Urban League.

In 1991, Tina moved to Sacramento, California as the inventory forester in the Region 5 regional office. She was responsible for remote sensing, photogrammetry and contract inventory for eleven forests, and did a lot of travel across the state.

A turning point in Tina's career came in 1994, when she took the position of the Tuskegee University Forest Liaison Officer in Tuskegee, Alabama. The program then was run by the National Forests in Alabama, but is now run by the Washington Office-Civil Rights Staff and the Business Operations-Deputy Chief area.

Tuskegee University was the first location that the Forest Service had ever established a liaison to recruit minorities into forestry. The school did not have a forestry degree and had not established a four-year degree program, but had arrangements for their students to get their forestry degrees from other schools. The students attended Tuskegee for two years and then continued at another school, usually for two years. Upon completing coursework, the student would receive a bachelor's degree from the other school. Thus, the student could never claim they graduated Tuskegee University.

Tuskegee is a small, rural town, and Tina did not want to go there initially; she had been raised to achieve, and her perception was that Tuskegee would not contribute to that goal. She talked with the dean, who became another mentor, about the program at Tuskegee. She was very impressed with the dean and Tuskegee after visiting the campus. During the interview, Tina identified an area for the Tuskegee University Forest Resources Program to succeed, and that pertained to expanding the program to have more degrees, and provide more students an avenue to be successful.

She told the dean and her supervisor, the forest supervisor, that she thought the school should expand a program called the 3+2 program. At the time the only university that was collaborating with Tuskegee to implement this program was Auburn University. The 3+2 program would allow Tuskegee to have a student attend two different universities to obtain two different degrees, a Bachelors' and Masters' degree. Tina also had a vision to expand the program so a student did not just have to choose from forestry or soil science, which were the only options available at the time, but to expand the program into other natural resource disciplines like ecology, wildlife biology and fisheries biology. The new liaison after the interview turned out to be Tina.

During her tenure Tina expanded the degree programs through eleven agreements with nine universities. She also changed the name of the program to the Tuskegee Forest/Natural Resources Program.

By 1996, Tina wanted to get back to her roots, doing land management. She wanted to be a line officer in the west. She got the job of district ranger on the Tonto Basin Ranger District on the Tonto National Forest in the very small town of Roosevelt, Arizona, two hours northeast of Phoenix. She fell in love with Arizona and the desert.

She chose to live in Payson, Arizona and commuted one hour to Roosevelt Lake. Tina has three criteria she uses to judge whether a town is big enough for her to live comfortably in: "it has to have a Safeway, a Wal-Mart or Target and a McDonalds."

Tina considers being a district ranger the best job in the agency; being on the ground dealing directly with issues. She had a big recreation program, with Roosevelt and Apache Lakes on her district. She learned that recreation should be managed more like the National Park Service recreation sites in some locations, and to manage recreation sites to fit with the image of "America's Great Outdoors." This means providing more amenities, such as flush toilets, sinks and soap and trash receptacles in recreation areas. Her motto was, "You really can't change people's mentality about urban values. People may be removed from an urban area when they leave their house, but they bring their urban values/ideas with them to the forest."

Tina left the Tonto in late 2000, to go back East. She loves politics and working in Washington, D.C. She had been through a leadership program in California called The Learning Exchange, and had made contacts in D.C. She received the job of a legislative affairs specialist working in the deputy chief's area of Programs and Legislation, which has now been moved directly under the chief's office. She is proud of that job, especially of having her fingerprints on eighteen pieces of legislation. Of nine people working in Legislative Affairs, Tina carried seventy percent of the workload in the National Forest System with responsibilities for the lands, range management, recreation and wilderness management and ecosystem planning staffs in the National Forest System. She also was responsible for all legislative issues in California, Arizona and New Mexico, and for the staffs in business operations: civil rights, Job Corps, acquisitions and human resources. She also had a part of the National Fire Plan in State and Private Forestry, and was responsible for all aspects of legislation pertaining to law enforcement.

Eventually Tina wanted to go west again, and in 2004, she took the forest supervisor position on the Cleveland National Forest in San Diego, California. It was a huge political area for community land management, and Tina's Washington Office experience had prepared her well for it. There had been a big fire on the Cleveland in 2003, and there were lots of fire issues. Tina had fifty-two fire districts to coordinate with around her forest, pertaining to over fifty-two communities. Her workload was huge and she was unable to get a deputy forest supervisor to help during her tenure.

In 2007, the deputy regional forester, Cecelia Bennett, informed Tina about a challenging position as forest supervisor on the Sequoia National Forest that she thought Tina was right for. The forest had a different set of complexities; there were six major issues/concerns that had been identified by the public and employees. There were lots of communication problems, between the public and the Forest Service, among employees, between forest and regional office staff and between district staff and forest staff. Lots of coordination and teamwork were needed. Tina's experience had shown her that change occurs with small groups committed to it, and that the process takes about 2 to 2½ years. Though she had not wanted to take a lateral assignment (vs. a promotion), she sacrificed her career goals at that time and accepted the challenge.

Regional forester, Randy Moore, suggested Tina look next to the Job Corps, which led to her current position as Director of the Job Corps in Denver, Colorado. President Johnson created Job Corps in 1964, in his campaign against poverty. The Corps was established to train poor kids, help them to earn their high school diploma or G.E.D. and to learn a trade. They are taught to function in society rather than just taking from it.

Tina feels that all her previous jobs and her training and experience in planning and structural process, has prepared her for this job. On the Sequoia she had learned to look at issues as a team with the public, have the public involved in managing public lands and have people involved in implementing projects on the ground. She says now that it is "all about logistics with a peace of mind; thinking about how to move from point A to your goal, how to get to where you want to be. Each manager should ask this question: How do you move stuff, people and issues — putting the pieces together to achieve results?" Thinking this way is part of Native American attributes that Tina identifies with since she is part Native American. She has a gift of vision and using it to solve problems.

When Tina got to the Job Corps, there were a number of vacant positions that needed to be filled, a task that normally could take a year or more. Tina led a team that filled over ninety jobs in one year (one project entailed filling twenty-two jobs in 3½ months — the quickest ever). She also working on developing processes and systems to work as a team and achieve results.

Tina loves the Job Corps work, but feels that the natural resource component is missing from her duties. Job Corps teaches about fire and forestry, but her position now is not involved in land management. She misses caring for the National Forest

System lands, working with the public. She loves the mission of the Forest Service.

Tina was also one of seventy people in the Senior Executive Service Candidate Development Program (SESCDP). The program is a unique program, overseen by the Office of Program Management, and involved an intense academic training to propel middle managers to become senior executives in the Federal government. The SES program came out of the changes that occurred in the civil service during the 1970's, where congress and the president identified a need to have senior executives to be trained to lead the federal government in serving the American public. There were sixteen people in her group (the SESCDP is broken into four different cohorts) three or four who had take-charge personalities, and Tina was one of them. As a result, she volunteered for a lot of extra work, typical of her self-proclaimed workaholic nature. In February 2014, Tina was promoted in place from Assistant Director of the Job Corps program to the SES position of National Director, Job Corps National Office.

Tina says her choice of a non-traditional career was probably deliberate. Counselors said she could be a doctor, lawyer or her pick of other traditional professions, but she did not want to be in a box or in a career in an office. She believes we all have unique gifts, and she does not want to be a robot. She said, "I want to get people to think, to take the road less travelled — or create a road where no road ever existed."

Every job has had challenges. One that sticks in Tina's mind was developing the new program from scratch at Tuskegee University that no one had thought of before.

Another challenge came when she was working in Legislative Affairs. Tina helped develop the Federal Lands Recreation Enhancement Act of 2005 (FLREA), and got it passed through congress. The act authorized federal agencies to charge a fee for providing a service, which was needed because appropriated dollars were not covering costs of maintaining recreation sites. In 1996, there had been a Recreation Fee Demonstration Project, a temporary, trial-and-error fee collection program that Tina had participated in on the Tonto National Forest. This Demonstration Project laid the foundation for FLREA. Tina set about creating a permanent program. This entailed drafting the legislation with her counterparts in the Department of the Interior and working with congressional committee staff. She worked with the Forest Service staffs and national forests, government attorneys and other agencies including Bureau of Land Management, National Park Service, Fish and Wildlife Service and Bureau of Reclamation, and with the Senate Appropriations and Natural Resources and House Appropriations, Agriculture and Resources committees. She also worked on and managed public involvement in crafting a bill working with non-profit organizations. She drafted the bill that went to a congressman. The Senate would not look at the bill until there was a hearing. The Senate had not had a hearing for over three years. Tina orchestrated a hearing by working with political appointees from the Departments of Agriculture and Interior who requested a hearing be held in the Senate. Her persistence resulted in the act being passed after working on the legislation for over three years.

Completing the *Giant Sequoia National Monument Plan* for the Sequoia National Forest was another great challenge. This project was actually Tina's biggest challenge, and also "my biggest success." When she arrived on the forest, she thought she could get the plan completed in two years. Late in 2007, it became obvious this was not going to occur. In fact, she had to change her focus, her vision and her mindset to accomplish this monumental goal. Before she left the Sequoia in March 2011, she was able to get the forest and the public to develop together a draft environmental impact statement and draft management plan for the Giant Sequoia National Monument. Along the way, she was able to develop many relationships in the community where other accomplishments have blossomed. One was the establishment of the Giant Sequoia National Monument Association, a non-profit friends group for the Giant Sequoia National Monument.

Another accomplishment was the development of numerous partnership agreements, challenge cost-share agreements, and volunteer agreements. A final accomplishment for Tina was to help the forest identify strategies and to develop and implement plans to complete projects. Tina took from her four years on the Sequoia "a breadth of knowledge, experience, humbleness and gratitude from the people I worked with and for." She once remarked to a newspaper reporter that she thought when she took the job on the Sequoia that she would change the forest. After four years, she realized she did change the forest, "but really the forest changed me, and it is because of the public and the community people, and the many, many relationships I developed along the way. I have changed and I am better for my time on the Sequoia."

Customer Service is Tina's greatest passion — helping assist the Forest Service in meeting people's needs. She also reminds us that she is a sports fanatic.

With a number of years left to work in government, Tina says it is too soon to decide on her greatest goal in life; she is happy about the milestones. She said, "I am intent on noticing and enjoying the journey and learning from it, rather than just focusing on achievement of a grand goal."

Tina currently lives alone in Denver, her twelfth location during her career so far. She has worked in fourteen different locations. She rents, since she moves and travels a lot. She did buy a condo when she lived in San Diego and in Visalia, California.

Family comes first for her, "even for a workaholic." Her mother is the most important person in her life. She sacrificed for her daughters, worked to make sure they had a better life. They grew up poor but not destitute, but sometimes they did not have money for heat or a lot of food. Her family never went hungry, but many times, they went with little. That taught Tina that the basics of life are food, clothing and shelter. She had food growing up and clothing and shelter, but her family was limited in the other comforts of life that people take for granted, "like heat, a bed of your own, vacations." She did not learn how to drive until she was twenty, did not have a color television until she bought one when she moved to California in 1989, did not take a shower or even knew what a shower was until she went to college in 1983. Now Tina and her sisters are successful in life, and "will take care of our mother forever." Their mother retired in 2002, and the three sisters pay her rent and other bills — whatever she needs, she gets.

Tina's first most extraordinary experience came early when she had to delay college because she needed to work to obtain funds to get out of high school and to help pay for her older sister's school costs. She realized then that she would not be able to go to college unless she took charge of her own life, and this shaped her life.

Tina said she finds it amazing looking back on life that every job or experience was due to something that happened before. She believes in fate. She finished with, "I am happy to enjoy the journey and people along the way. I have a boundless network."

Chapter 8
Line Officers

Line officer positions in the Forest Service are positions of leadership. The women featured in this chapter sport lengthy careers in the Forest Service. All have held multiple positions on their way up, some even crossing disciplines, so this chapter further showcases the variety of positions available in the Forest Service and the different pathways taken by these women to become line officers.

Grouped here are line positions at the district, forest, regional and Washington Office levels. The roles of district and forest levels have been described in previous chapters. Regional office staffs coordinate activities between national forests and grasslands, monitor activities on those lands, provide guidance for forest plans, and allocate budgets to the forests. A regional forester oversees forest supervisors. Regional foresters report directly to the Chief, a federal employee who reports to the Under Secretary for Natural Resources and Environment, U.S. Department of Agriculture.[49] The Chief's staff provides broad policy and direction for the agency, works with the President's administration to develop a budget to submit to Congress, reports to Congress on accomplishments and monitors activities of the agency.

One notable common characteristic among the women featured in this chapter, at any level, is their love for and devotion to the land. For them their work with the land and its people is a means to accomplishing their goal of leaving the world a better place. A line is crossed from field-oriented duties into line duties in order to advance into these positions, which can be a struggle for some. The district level remains closest to the ground, and while duties at upper levels mean a lot of working indoors, the objective is always caring for the land while serving people.

Rachel Feigley, District Ranger

Rachel's family moved from Missoula, Montana where she was born, while she was a baby. She grew up in Nebraska in Scottsbluff, Omaha and Lincoln. She lived in the city through young adulthood, but said she never felt she belonged there. She was regularly exposed to "infusions of outdoor life," which were the largest influence on where she ended up living and the kind of work she chose. Once introduced to this alternative, she was always drawn to rural and more natural settings.

As a child, Rachel (Figure 141) liked playing anything outside. She said, "I certainly went through a tomboy stage but always had a group of girlfriends. I played kick ball and cowboys and Indians but also loved Barbie dolls and slumber parties." She also liked organized activities with girl scouts, church camp, pep club, choirs and choral groups and leadership cadres. She disliked bullies and organized sports.

Figure 141. Rachel Feigley (right) and husband, Pete, at Black Mountain, Absaroka Mountains, Montana.

Her most memorable and cherished values were family vacations. Her dad was a Lutheran pastor, always ministering to his congregation. During their vacations to Swan Lake, Montana, they were "a family all to ourselves." Her family of six loaded up the station wagon, drove 1,500 miles and rented a cabin for a couple weeks to fish, boat, water ski, swim, pick berries, watch sunsets and roast wieners and marshmallows.

Rachel's favorite school subjects were math, Spanish (which she still uses when traveling), biology and chemistry. At the risk of "being a nerd," she was on the math team for a short time in junior high. She said, "I always loved math. Science was also interesting to me although it was difficult to separate all the physical sciences from the biological. The two things that really clicked for me were taking biochemistry, which brought both physical and biological sciences together, and a lecture I heard that described a pyramid of all the sciences, and math was the foundation. Everything was built on everything else and math was the language that helped explain it all."

At fifteen, her mom encouraged her to compete for an Isaak Walton League grant that sent her to work on a Youth Conservation Corps (YCC) job crew for the summer. She was introduced to manual labor, using tools, teamwork, camping and potential outdoor careers. Her crew dug outhouse holes, installed parking posts, built a sod house, picked up litter and constructed retaining walls in several State parks and communities.

After her YCC experience, Rachel always looked for jobs outdoors. She knew about the Forest Service since her brother had a forestry degree and worked for the agency. During college she worked summers for the Forest Service as a biological aid. She said, "I started out as a GS-2!" Before getting her permanent position, she had worked ten seasons at several ranger districts in Utah and Montana.

During the winters at college, Rachel was an officer in the Wildlife Club and Range Club and was on a team that competed in the Society for Range Management annual meeting plant identification contests. It was clear that she wanted to pursue a career in natural resources, to become a steward and to continue to feed her "new habit."

Rachel said she did not deliberately choose a non-traditional career, "I just followed my heart." She never noticed being a minority in her classes — she said she thought her generation had begun to change that. She did notice being a minority when she started working for the Forest Service in 1978. Men then held most seasonal jobs. Rachel was the only female as a seasonal on district crews, and once permanent, one of few non-clerical, professional women. She often worked alone or on a crew of mostly men, but never felt that she did not belong. There were times she felt like she was being observed for her overall tenacity, skill and ability. She worked hard and felt she gained respect, and had positive experiences with most of the men she worked with. She said, "I was fortunate to work with men who, once they saw what a hard worker I was, accepted me, but also kind of took care of me." She started seeing a shift by the mid-1980s, to more women in permanent positions other than administrative or clerical positions.

As many have faced, it was a challenge for Rachel to get a permanent position with the Forest Service. She had passed up offers to "fast track" through special programs, as she did not want to lose her flexibility to choose where to live and work. She had been a seasonal for years and had given up and gone to the outfitter guiding business. A window opened in 1990, and she got her first permanent appointment with the Forest Service. For the early part of her professional career she was a range conservationist. The majority of her career was spent as a district wildlife biologist, and until recently she thought she had found the ultimate career in that position.

As a biologist, Rachel's workday varied. Her calendar filled up quickly with meetings, mostly facilitating communication between resource specialists and managers, community members and/or interagency partners for project planning and implementation. She kept up with the latest research on lynx, wolverine, grizzly bear, bison, elk, etc. Days in the field were at a premium and involved supervising crew work, monitoring

Figure 142. Rachel, bird in hand on Migratory Bird Day.

implemented projects and collecting new information (Figure 142). Field work included conducting goshawk surveys, aspen inventory, forest carnivore snow-track surveys, project level habitat reconnaissance and the like, based on interagency agreements or the "issue-du-jour (Figure 143)."

Rachel's position morphed during her years as wildlife biologist in Livingston. Due to budget trends, the Gallatin consolidated districts and created zones to share employees, and eventually the consolidation of the Custer Gallatin Forest left one biologist where there had been four. Rachel had responsibility for the Absaroka-Beartooth 'zone' of the Gallatin, totaling 1.5 million acres. This was a breaking point for her — professionally she felt that she could not effectively serve in her role and personally she had to lower expectations of what she could accomplish. It inspired her to look for options. At the risk of giving up the dream job in favor of not burning out, she sought out acting district ranger details and applied to locations that suited her.

Figure 143. Rachel birding on Migratory Bird Day.

Rachel eventually became the district ranger very near the location of her childhood family vacations. She was thrilled to live and work in this gem of a landscape. The change from resource specialist to district ranger has been both challenging and refreshing. Being a line officer responsible for a variety of disciplines requires being attentive to a fast learning curve and relying on experienced staff to navigate through decisions from simple permits to complex land management issues. Rachel said, "While I thought myself fairly well experienced on a variety of disciplines, either from having to respond via NEPA or having had some program manager responsibility role, it was not enough to prepare me for all the things that arrive on my desk every day. Issues related to lands, special use permits, supervising personnel, partnership agreements, etc. I am thankful every day for the knowledge and experience I do have in the resource arena which seems comfortable and does not require the steep learning curve." Although she spends much more time in the office than she would like, she feels that she has the opportunity to make a difference. She said, "I am so happy to be in this beautiful landscape working with a great staff and awesome partners who are passionate about our national forests. It is truly a privilege that sinks in a little more each day."

Being successful in her job is important to Rachel. She derives a lot of satisfaction and pride from her work, but believes that work is a part of her life, not her whole life. She said her biggest challenge has been balancing the love of her job with personal dreams of home and family. She works hard and plays hard. Her greatest passion is our natural resources whether expressed in her job as a wildlife biologist or district ranger, her lifestyle, or her involvement in community and world natural resource issues.

She is still working on defining her greatest goal in life. She said, "There are so many!"

Coffee and yoga with meditation are an important start to most of her days. She feels she has been blessed and has faced no significant challenges other than those we all face in life-loss of friends and family. She said that working in small towns or remote locations was not very conducive for finding eligible men, but she loved that lifestyle.

She is now happily married to Pete, a biologist and musician, whom she adores. They currently live on the ranger district compound, renting a government-owned house that affords ready access to both the office and the woods. Her family is relatively close and they get together for most holidays and special events.

Extraordinary experiences have abounded throughout Rachel's career. One of her seasonal positions was as a wilderness guard. She spent the summer hiking, backpacking and horse packing into various wilderness locations on the Hebgen Lake Ranger District in West Yellowstone, just west of Yellowstone National Park. She had the privilege of living at the historical Basin Station, which is now a Forest Service rental cabin, near Hebgen Lake. The job entailed visiting with wilderness users, conducting campsite inventory, packing out garbage — "not glamorous work but the backdrop was fabulous!" Those were awesome days and nights; being in touch with the moon cycles and planet alignments, with wildlife movements and the seasonal changes was uniquely satisfying.

In 1995, after moving to her second permanent location, Rachel took a detail to Alaska's Kenai Peninsula. It was a dream detail. She worked with two field crews, an alpine crew and a forest valley crew who were sampling vegetation for the development of a vegetation classification. Flying into remote areas for days at a time, hiking past blueberry and salmon berry bushes with berries the size of her thumb, seeing grizzly bears, caribou and moose up close, seeing the northern lights in the middle of summer, hiking after work until midnight and packing a gun with 500 gram slugs, of which she said, "Don't quote me on that–they were shotguns with huge slugs." And the long days of work meant they got to enjoy three-day weekends.

Rachel sums up her career experience by commenting, "Being a Forest Service employee is an honor and a privilege. Being a woman in the Forest Service has been an exciting and demanding challenge. Some of my life's greatest role models are Forest Service women! We have helped shape the new face of the Forest Service and served as strong resource advocates."

∽

Molly Fuller, District Ranger

Molly's mother raised seven children, after her dad passed away when she was only six years old. Growing up, she had an incredible amount of freedom. She was the youngest, and her brothers and sisters generally kept an eye on her, "but not like people do today." She was attracted early on to what kids called "the fields," which were the undeveloped estuaries and marshlands of the San Francisco Bay, on the California coast. She said, "I'd only have to hop two fences and cross one railroad track to get out there, but once there, it was me and nature. That was my escape, my hang out and my earliest years. It was, in retrospect, a safe place, yet kids today fear these places

Figure 144. Molly Fuller in Emigrant Wilderness above Big Lake.

the most. These wanderings in my early years influenced my life passion and career as a conservationist."

Growing up Molly (Figure 144) found her "industrious vein." For 1½ years she was a babysitter for a family of six. The kids were ages six months through eight years, and Molly was thirteen. She took a second job at the high school during the summer operating an old-fashioned switchboard, one of which she marveled she "recently saw at a museum!" She also worked a summer at the Air National Guard, as a clerical aid for Sgt. Wolfgang Klamp. Of that job she quipped, "I was sixteen, my sister Nancy taught me to drive and I had a boss named 'woof.' I always felt like I was barking in that office, as he was sometimes hard of hearing. But I was becoming more and more independent."

During her senior year of high school she took a vocational course "Project Invest," a simulated insurance agency for all practical purposes. At that point, all she wanted to be was "just like my mom," who was a legal secretary. She learned to type fast and accurately, and got a job with an insurance agency. All summer long she sat at a desk, in the mandatory panty hose and polyester suit. Working full time after high school, she found herself staring out the window, wondering about her future. She realized that she did not want to be inside behind a desk for the rest of her life. She respected her coworkers, two middle-aged women who had resigned themselves as clerical aids in an insurance agency, while the agents "enjoyed all the spoils of the business." Those women taught her well, "gave me the courage to explore." Molly's mother fought hard to convince her to go to college, and was relieved when Molly announced her decision to do so. Molly said, "She was my greatest support that propelled me to college."

She chose a college that she could afford. It was too far from home for her family to just drop in, but close enough to keep in touch. Instinctually, she chose agriculture as a major. She also wanted at least one person near that she knew would help her if she got in trouble. She chose Chico State College in Chico, California. Her lifeline was Pattie Gouveia. She was a friend of Molly's sister, who adopted Molly in college. She hired her to help her graft Kiwi's. They worked in the fields with Mexican men and women, doing the hardest work our "modern" society still refuses to understand. The farmers were transitioning orchards from wine grapes to kiwis, an experimental thing in the late 1970s. It was a specialized skill and they were paid $2.00 per graft. It was her first "field job." It was incredibly hard work, but they made what they thought was very good money.

During her junior year Molly heard of a job for the Forest Service, and she still has the resume she sent to the Eldorado National Forest. She was hired in 1980, in a cooperative education program. She worked for the natural resource officer in Placerville, Eddie Rael. He was from New Mexico, catholic, had seven kids — "a father figure." He was a man of color, highly respected and he represented diversity. Molly loved him. He died six years into her career on the Eldorado.

Eddie had had an interesting and eclectic group of "specialists" that he had nurtured. Anne Denton, a MBA graduate from Stanford, now a district ranger on the Stanislaus National Forest, KJ Oberman, a soil scientist (now KJ Silverman), later a deputy forest supervisor on the Mount Hood National Forest and Judie Tartaglia, a biologist and eventually deputy forest supervisor on the Tahoe National Forest, and now retired. Molly said, "These are the women who helped me navigate the predominantly white male organization. These were the pioneers who saw the justice, the injustice, the equality and the inequality. These are the women who had a voice and made their voices heard and who helped me find my own." And she remembers many others fondly: Karl Stein, who started when Molly did as a cooperative education student in fisheries, who worked for Jeff Kirshner and Mike Henry, forest fish biologists. Chuck Mitchell, the forest soil scientist, Mike Kuehn, the hydrologist. Charly Price, who was an artist and is as far as Molly knows, the only professional illustrator in the agency and has had a huge impact on western conservation education, and the "taciturn and enigmatic" Diana Grettenburg who was a landscape architect and Molly's roommate one summer. And there was Lou Merzarrio, who was the flashy, passionate planner, and Jess Barton, who was a patriarch in many ways in his role as the planning officer for the forest. Molly remarked:

> These are the people that helped shape my career and my life. I babysat Chucks kids; we had a forest softball team "the banana slugs," I learned to fly fish and made my first fly rod in Jeff K's garage, Mike Henry loaned me his baby crib, I broke my toe working for Mike Kuehn, who named a meadow after me. These people helped me make decisions, big-life career decisions that have had huge impacts on my life. My first was my decision to make a living as a conservationist. That public land conservation was worth all the slings and arrows, all the hard dirty work, all the drudgery and grief. I learned from these early years my passion for natural resources, the importance of scientific curiosity, and the foundation of scientific principles applied to public land management decisions. They encouraged me to find a detail or job on a ranger district, which I did.

Figure 145. Molly on horseback.

Molly worked for a resource officer, Rich Platt in 1984. Their district ranger, Craig Harrasak, hired her full time to be Rich's assistant. Rich was short, and dyslexic. Molly was tall and left handed. They did everything together, a "couple" of sorts, joined at the hip. Rich said he taught Molly everything he knew. How to tie a knot, which knot to tie, how to saddle a horse (Figure 145), pack a mule, how to respect the public and know when you're getting swindled or when someone was pulling your leg. She said, "He understood and shared the notion of public service. He was inclusive in everything; you just needed to show up. He was my friend and my supervisor. He didn't favor anyone, but allowed you to share your passions, your vulnerabilities."

By that time Molly had been married to her husband Bill for about two years, and she was pregnant. She had a baby boy, Elliott, and went back to work before she weaned him, still a nursing mom. She held her son while counting cows onto the Pacific Ranger District (Figure 146). She has memories of being in the field with Rich and asking him to stop the truck because she had to "pump" her breasts. She said, "He was so tickled, you knew he loved me. He stopped without hesitation, gave me all the time I needed and we continued on. He was always game." Molly broke her leg on a day off. It was a major accident causing nerve damage, and she was in the hospital for fourteen days with high potential to lose her leg or the functioning of her leg and foot. By that time Elliot was eighteen months old. Molly said, "It was the singular worst accident in my life, big life event. But the Forest Service saw me through. Rich didn't miss a beat, he was there and he got me back in the field. I remember I still had the cast on. He laughed at the 'foot prints in the trail,' which consisted of my normal shoe, the cast and the cane. I often think how I might have given up on ever walking normally again if it hadn't been for Rich Platt. But walk again I did. And I skied again and everything else."

Figure 146. Molly nursing Eliot while working the range.

Molly said she learned about harassment, first-hand, and more importantly, she learned how to deal with it early on. She had strength, faith in herself and her coworkers. Her first harassment experience was with an older guy named "Dick" (how appropriate, she commented). She was twenty-two and new to the agency, and Dick invited her out for drinks with other coworkers after work. Sometimes she found it was just her and Dick, and it got fairly "chummy." When she often found herself alone with him, she thought it was odd. Then the old-timer dispatcher, Ed Grosch, took her aside and told her that Dick was an asshole and had a tendency to "pick-up" unsuspecting women, despite being married. Molly was floored. Ed had barely spoken to her before, and he got her attention. She stayed far away from Dick after that. Several years later, a friend confided in her that Dick had done the same thing to her, only worse. They exchanged stories, and discovered others had fallen "prey" to this guy. Eventually a grievance was filed. All Molly knows is that Dick disappeared. She said, "Amazing. It seemed like the system worked. No fan fair, no courts; but justice was served. What it took was good people, willing to speak up, willing to work together. I think of this as a win-win."

Figure 147. Molly — range conservationist in uniform.

Molly spoke about the consent decree, the class action lawsuit that changed the Forest Service in Region 5:

> It became, in my perception, a court mandated bureaucracy. "A terrible punishment for a terrible crime or what I think of as a 'lose-lose.' I'm not sure which is worse to this day, the crime or the punishment. I know others who experienced the blows of inequality and injustice. But there was an important network of people, both men and women we could trust and get support and seek justice. It was that simple. It had nothing to do with the consent decree, and I firmly believe that the consent decree, to this day has done more damage to good people than it has improved working conditions or fairness or justice. I often think that the Region 5 consent decree was like a watered down version of 'McCarthy era' but instead of fighting communism, we were fighting harassment and prejudice. It just seemed to become this negative, over-blown, pretentious bureaucracy. There may have been good intentions, but that is where it ended. How could it have been different?

Molly's first big promotion and move was to the Mendocino National Forest (Figure 147). District Ranger Tom Mainwarring recruited her. He said she had "moxie." She was pregnant again, and when she broke the news, he did not skip a beat. The Mendocino was supportive of working moms (Figure 148). Molly exclaimed, "Unbelievable! This was back in 1989. They set me up with a computer at home. I could take as much leave as I needed, and I did. I had my baby girl, nursed her as long as I needed, kept up with work and people just kept helping me out. I can remember going to a training session on Wilderness up at Beaver Glade Work center. I took my baby Alexis. We stayed in our van. It was a big camp out and we hiked into the Yolla Bolly. Everyone seemed to enjoy having us along. I guess looking back . . . I did have moxie!"

Molly spent ten years on the Mendocino, the longest she has stayed on any forest. She raised her kids in Chico for the most part, while she worked at the district. They went through downsizing there in the mid 90s. Molly lost many fellow employees to reassignments and retirements. People

Outdoor Women inside the Forest Service

Figure 148. Molly with kids at work.

had to recreate themselves in many ways. Some struggled, others found opportunity. Molly fell into the latter category. She collaborated with the law enforcement officer and they purchased two horses for the district that became part of their staff. Molly said, "I got a lot of shit from people for doing that. But I'll never regret it. It kept us busy and in the end, everyone loved those horses. They got us just about everywhere; kept me out of trouble. At the time, these horses were the best public relations staff the Corning Ranger District had, both internal and external."

Molly said she had become a fairly decent fire fighter, eventually a crew boss and a field observer in the fire organization (Figure 149). This was exciting to her and gave her a huge boost in confidence.

Line Officers

Figure 149. Molly's postcards sent home to kids while she was away fighting fires.

She said, "I owe the Mendocino so much for this part of my career." She was also able to develop ecosystem skills in restoration work (Figures 150 and 151). The staff on the forest was "incredibly future minded."

The forest was the southernmost range of the Northern Spotted Owl. It was part of President Clinton's plan introducing "late successional reserves" to provide a habitat corridor that would help insure the viability of species associated with the habitats of late successional ecosystems. Local communities dependent on the public land policies that encouraged multiple uses were struggling to downsize and detach. Mills started shutting down along the west coast. Economies were turned upside down at an alarming rate. But according to Molly, the forest staff managed to do the right thing. "They maintained a focus from an ecosystem perspective. Way ahead of the curve in terms of fuels management, in terms of thinking about community and human interactions. I learned so much about how it all fits together, as painful as it was. I look back with great fondness of the Mendocino; the staff at the time was patient, supportive and tolerant and demonstrated a huge amount of foresight. I think the way the forest looks and feels today is a testament to the many who came before."

She was struggling a lot with her personal life in the late 1990s. After the districts were consolidated her supervisor was District Ranger Jim Giachino. Molly applied for a "sabbatical" one summer, and he gave her "full support to figure myself out." Her leave of absence was fairly open ended. During that time that she decided that she needed to do something different, "or resign myself for the next twenty years in the same spot." She applied for a planner position with the City of Chico, an environmental restoration supervisor with Nature Conservancy on the Sacramento River and a job with the local school district in Butte County. She was interviewed for two of the jobs, and The Nature Conservancy offered her theirs. It was a big decision for her at that time, but she turned them down. She said, "I often think where and what I'd be doing if I'd said

Line Officers

Figure 150. Molly checking stream condition.

Figure 151. Molly inspecting trees.

'yes.' Then I applied for a job on the Plumas National Forest, and lo and behold I got the job."

She went to the Plumas in 1999, as an ecosystem manager. She did not know what that job could possibly entail, which is why she had applied. She figured she would be able to make it up. The agency was headed into a new way of thinking. Molly said, "We were turning a huge corner in public land management policy and I felt like I was on the cutting edge. I was part of a 'team' of ecosystem managers, one on each district, and a staff officer at headquarters in Quincy. We were implementing the Herger-Feinstein Quincy Library Group Forest Recovery Act (HFQLG)."

The changes happening on the Mendocino under the president's plan were also happening on the Plumas. The big difference was the HFQLG was a grassroots movement. Locals had convinced a republican representative (Herger), and a Democratic Senator (Feinstein), to collaborate on a land management approach to fire adapted ecosystems in the northern Sierra. They passed a bill in 1998. It was an ambitious idea, the pace too fast and scale too large for the bureaucracy to handle. There were lots of reasons it could not be done in the four-year term of a presidency. Molly spent eight years there, and it seemed by the time she left that they had merely nibbled around the edges of implementing the HFQLG Forest Recovery Act. She observed:

> Nature has a way of making you feel very small. The predictability of congress or budgets was no comparison to a wildfire, flood or other catastrophes. Looking back I think that HFQLG demonstrates the beauty in the inefficiencies of the federal government. It was a grand 'pilot' project, to implement fuel treatment projects on not less than 40,000 acres per year. It was supposed to be done for five years, and we would study it to see if it was a good idea. So by 2005, we were supposed to be done, have treated 200,000 acres with strategic fuel treatments, oh yes, it was going to in part pay for itself with the materials we were treating. Also we were going to test the efficacy of group selection timber harvest, which was also going to help pay for the project. And why

not? The public inholdings of the pilot project was 1.53 million acres, you'd think it could be done. Plus, this idea was going to work because everyone was holding hands in the community and we had an act of congress to back us up. Now something like this could happen in Russia, or Brazil, maybe even parts of Mexico. Might even be a good idea on private land in the US. But on federal land in the US? Not! Wally [Herger] and Diane [Feinstein] needed to have modified all the other laws that applied to federal land management decisions, even needed to 'uncomplicate' the constitutional rights of Americans to file a lawsuit to make this happen in five years. I was there for eight years. I consider each year was like a dog year, it was worth seven anywhere else. So much pressure, almost unbearable. Not to mention the egos that you'd be dealing with (including my own precious egosystem!).

Molly thought the best part of the ecosystem manager (EM) job on the Plumas was being on a team. The EMs met once a month and commiserated, planned, joked, cried and told unbelievable stories. They were dealing with huge budgets, understaffed organizations and personnel issues — it seemed like the stack was against this pilot project and it never occurred to them that it was doomed from the day it was signed to meet the five year commitment. After five years, Feinstein and Herger "got the damn thing extended!" Meanwhile, fires kept happening, along with other disasters, including 9-11, the challenger crash, Katrina and the list goes on. The ecosystem managers stuck together, through thick and thin. Molly says, "It was a bond made out of pure adrenaline and fear. You had to laugh and you had to work together. One failure in sticking together amplified failures across the landscape. It really demonstrated the need for teamwork. We were affectionately called the 'auntie Ems' (referring to the character in the *Wizard of Oz*) because the three of us were all women of the same cohort, in our mid to late 40s."

At one point the "aunt-ems" got sick of meeting once a month in office conference rooms and arranged a meeting at Bucks Lake, which was central to the three ranger districts. As a joke, Molly called the meeting the "Bucks Lake Summit." The name stuck, and this annual meeting became the most important, popular meeting on the forest. Molly said, "I'll never forget the power of a team, and I miss my 'aunties' now. It is hard to replace the friendship and partnership that we created, not to mention the power of pulling together."

Molly and her husband raised their two kids in Quincy. Bill spent seven years in Quincy running Monolith Music, selling musical instruments. Both kids went to Quincy Senior High School. Elliot went off to college in 2003, and then when Alexis was a senior, Molly started looking for a new job. She got the U.S. map out, the one that showed all the 156 national forests. She put pins in the places she wanted to go, and applied when jobs came up. She was interested in district ranger jobs; felt it was a natural for her. Bill could see the economy was not going to favor a small musical instrument store and he was tired. He had run that business seven days a week, for long hours to make it work. He was supportive of closing the shop and exploring the world with Molly.

When Molly was offered a ranger position on the White Mountain National Forest, over 2,500 miles away, it was a chance for adventure that she and Bill could not resist. When they had been there for four years, the Forest Service just celebrated 100 years of the Weeks Act, which underlined the importance of public lands and conservation. In the west, this country carved out national forests and parks from "public domain," lands that were purchased or won by the U.S. government, but had never been privately owned by anyone. In the east, all the land had been sold, either by the king, or the U.S. once the constitution was signed. There was no public land out there, just private industrial forest land, until 1911, and the passing of the Weeks Act. Molly said, "It is amazing how people here will tell you that their organization has been here longer than the forest. They have a story to tell. I love hearing it; I do not get tired of it. I listen to the partners, the history, unfolding of the story to present day. I use these stories, this history, to make decisions about tomorrow. It is really sweet here. My job is unfettered with all the craziness of the west." She said there were some issues, but they seemed manageable. People agree with what the Forest Service does for the most part. They have respect for the federal government. They may not agree with it all, may think the Forest Service is slow and has burdensome processes. But overall, there seems to be consensus and support. The Land and Resource Management Plan got approved in 2005 without appeal, which Molly thinks was unprecedented. She said, "I am amazed and delighted every day by the high quality of employees, the devotion to the concept of conservation and doing the right thing. I really got into a fantastic job here and a landscape that is completely new, not to mention a culture that is unfamiliar. People have a dialect here; they have a way that isn't what I'm accustomed to. It is what happens when you move so far away. I've been able to visit (now many times), NYC, DC, Boston, Quebec, Nova Scotia and parts of Vermont, Massachusetts, Connecticut and Rhode Island. It is truly an adventure. But I'm not done, I have a way to go."

Molly opened our interview by saying, "I am an extraordinary woman living my dreams. At least that is what I say to myself whenever I'm unsure about what I'm doing, or need the courage to do really big things." Her reflections on her life and career are best summed up in her own words:

> Am I an extraordinary woman living my dreams? I say that to myself because it is the hardest thing I can conceive about myself. The fact is, I am terribly ordinary. I pay attention to my dreams, but I have no control over them. But I am convinced that my dreams do help me understand who I am, they help guide me. I'd like to think they make my adventuresome self stand out.
>
> I think how I grew up with my sisters and brothers with the leadership of our mom, how independent I was. I got married to Bill early by today's standard (twenty-four), had two children, raised them, all the time working in these incredible settings of the national forests. These are the things that have shaped my life. This is what makes me the best that I can be.

I think of the excitement of getting up early to ride the range with the grazing permitee, or plant trees with the crew all day, or to call owls or measure stream widths or fish. I think of all the times I got up early to meet up with the fire crew, to do a prescribed burn, or starting late in the day by picking up a fire crew to head out for a fire. Or the times I've chased a smoke to put out a fire or the times I've called in a smoke to the dispatch. I remember all the helicopter trips (not just on fires), the long backpack trips into the remote Mokelumne, Desolation, Yolla Bolly, Snow Mountain, Bucks Lake and Pemigewasset wilderness areas. I have spent ten days by myself (if you don't count my horse), packing down trails, bucking up the trees, counting the cows, moving the cows. It was endlessly fascinating, exhausting. All the reconnaissance trips, or the conservation education programs, the living history costumes and or the campground programs. Or the times I've had to evacuate people, ask them to move for their own safety. Tell people, no they can't do that, or encourage them to try something new. Of all the times someone wanted something that they couldn't have, or tried to swindle the public; how you didn't let that happen (not without a fight), and the agency backed you up. In closing this . . . I don't think I'd change a thing and can't wait to write the next page in my bio.

Molly left New Hampshire a few years ago, finishing up her career back in the West, as the district ranger on the Summit Ranger District of the Stanislaus National Forest. Again she made her mark, leading her district team to receive the Region 5 District of the Year Award in 2016. She retired in 2017, returning to her home in Chico, California.

~

Beth Humphrey, District Ranger

Beth (Figure 152) grew up in Albuquerque, New Mexico. She spent most of her time playing in her yard, barefoot and covered with bruises and cuts, building things, catching bugs, toads and baby birds. When she later worked for the Forest Service, a rancher who asked where she had grown up remarked, "You may have grown up in the city, but from

Figure 152. Beth Humphrey.

where I stand, the city is gone from you and you have become a country girl."

She does not remember playing with dolls very much as a child, or playing house, but had stuffed animals instead. She was not interested in typical girls' activities. She played in the dirt, worked in the yard and hung out outside all the time. She was shy, but there were lots of kids her age on her street and they played together all the time. She climbed the trees in her yard and found birds nests with eggs in them. She said, "That was the best thing ever." She also made a habitat for "roly polys" (sow bugs) in her little red wagon. She put dirt and rocks and plants in the wagon, then caught the bugs and put them in. She said she always wanted them to have babies and she learned that if they were yellow on their bellies, they had eggs. One of the neighbor kids would go out on the mesa and bring back tadpoles and bull snakes. Beth's mom bought a five-gallon aquarium and they raised the tadpoles until they were toads. She said, "It was all about the animals for me. I loved my dog Tippy and my cat Copper. My neighbor had pigeons and we spent hours in the pigeon coop. I made a bird feeder using my dad's tools and scrap wood. I cut my big toe with the saw, but finished that bird feeder and put it in the mulberry tree outside my bedroom window."

Beth was athletic and loved playing sports and riding her avocado green banana seat bike with her friends, racing up and down the street. Basketball, tetherball, Ping-Pong, tennis, softball and volleyball were her favorites, especially basketball with the neighbor boys. They played HORSE and half court basketball in a neighbor's driveway. She always had great male and female friends, but gravitated more towards boys. She said, "It seems I had more in common with boys, felt much more comfortable and had an easier time connecting with boys. Even now, some of my best friends are men."

She was on sports teams in junior high and started tennis in high school. She was a tetherball champion in elementary school, a paddleball and Ping-Pong champion in junior high and played volleyball and softball on junior high teams. She played on a co-ed softball team in college, and played a lot of volleyball and tennis with friends.

During college she discovered fishing. She said, "I would drive for hours to go fishing. I had no idea how to fish, but I taught myself and still fish to this day."

When her dad was still alive Beth's family spent most weekends on the road going to forests in northern New Mexico and southern Colorado. They would picnic and play in streams and rivers. They never camped due to her dad's health, but they were always out in the woods on day trips and even overnight trips to Durango, Ouray and Silverton, Colorado. Beth remembers spending quite a lot of time at the sulphur hot springs in the Jemez Mountains. Since they lived in Albuquerque, they also spent a lot of time in the nearby Sandia Mountains just east of Albuquerque.

Beth's dad instilled in her a love of gardening and growing things. They had a big garden in their backyard with fruit trees and grapevines. She followed her dad around while he planted and harvested the fruits and vegetables. He planted a beautiful lawn and she helped him mow it. She loved her family but was closest to her dad.

Her favorite school subjects were physical education, spelling and writing and science. She was the school spelling bee champion in sixth grade and was in a regional spelling bee in seventh grade. She did not make it to the national finals, but was always good at spelling. She liked science most of her life, but was lousy at math early on. She said, "In the fourth grade, they tried to teach me fractions. I just didn't get it. Finally, they showed me a pie cut into slices, and that's when I figured it out. I guess I needed a visual. After that, I did well in math. I particularly liked geometry and solving word problems."

She said, "And then there was band. I started playing the clarinet in sixth grade. I was first chair clarinet in junior high and progressed into symphony band in high school. I discovered marching band in tenth grade and spent the rest of high school on the flag corps. We were state champions several years in a row and traveled to contests and shows in New Mexico, Arizona and Colorado. Those were great times. At New Mexico State University I was in the Pride Marching Band flag corps for four years, and instructed the last year or two. That's how I chose my university; I wanted to march with the Pride."

Beth received her Bachelor of Science in Agriculture with a Wildlife Science Major and Range Science Option from New Mexico State University in 1989. She was in college on and off between 1980 and 1989. At first she majored in biology intending to go into pre-veterinary science. She said, "That lasted about a year and a half, until my grades indicated I would never get into veterinary school. From there I floated from biology to geology, then landed in the wildlife science program." She was a minority once she started working on her degree in wildlife science. She said, "If I remember correctly, of the 200 or so students in that program, there were twenty-six women. I don't remember thinking I was a minority while I was in biology and geology classes."

She did not deliberately choose a non-traditional career path; there was nothing deliberate about what she has done with the Forest Service. While she was in college she worked summers for the Forest Service, and was happy to get a job, any job. It just so happened that the job she started in was as a seasonal range technician in Mountainair, New Mexico on the Cibola National Forest, an almost exclusively male-dominated job even now. She said, "I fell in love with the work and the Forest Service. I worked hard and my supervisor liked me, so I got to come back for a second season." She later hit a couple of rough spots with other supervisors who discriminated against her, sometimes subtly, sometimes blatantly. She did not let it stop her from doing the work she loved.

On her way to her current position as district ranger, Beth worked in numerous locations in New Mexico, Colorado and Arizona. She has been a range technician, a forestry technician where she picked up experience in fire fighting and silviculture and then she held progressively more responsible positions as a professional range conservationist and a wildlife biologist at various levels of the organization.

Her season as a forestry technician was a true "trial by fire," being Beth's first encounter with a supervisor who made her feel as though she did not belong. It took her awhile to realize why she felt as she did because she had always been shy and could be nervous around people. She talked about this instance of feeling she did not fit in:

> I was working as a fire fighter on an engine crew in Magdalena, New Mexico. I worked for a Hispanic male in his fifties, and was part of an otherwise all-male crew consisting of three Native Americans, three Hispanics, and me.
>
> The crew boss was not allowing me to work with the crew and made me ride with him when we went into the field. He told me the crew was uncomfortable working around me. I didn't really understand why they would be uncomfortable since I was rarely with them.
>
> One day, the crew boss and I were out in the field alone checking out a fire that had burned previously and he told me I needed to tell him I respected him. Being a young

woman with a smart mouth, I told him I wouldn't tell him that until it was true. He told me to tell him I respected him. I said no. He stopped the vehicle and told me to tell him I respected him. I told him no and asked him to keep going or I was going to get out of the vehicle. I got out of the vehicle and started walking. He came after me and quit bugging me and we went on.

After that, he went on vacation for a week or two and I rode with the rest of the crew. They asked me why I didn't ever work with them. I told them I was told that they were uncomfortable working with me and that's why I always rode with the crew boss. They told me the crew boss had told them I didn't like them. We all became friends and worked well together until the crew boss came back. At that time I was working with another engine crew getting a float ready for the 4th of July parade. I was talking with another crewmember that was telling me of difficulties he had with my crew boss. The crew boss overheard and came up to me and yelled, "Do you want to fight me? Come on, fight me!" I told him I didn't want to fight him. The next day I was called into the assistant fire management officer's office and reassigned to the timber staff due to my being a problem.

Following that season, Beth entered a cooperative education program, combining related seasonal work with her university program. It led to a permanent position following graduation. During her first season of range work in that program, she again experienced a sense of being a token and not belonging, which she described:

This instance occurred with a different male supervisor when I was a cooperative education range conservationist on the Mountainair Ranger District on the Cibola National Forest. I had been on the district under a different supervisor for two seasons, skipped a season, fought fire and then came back as a coop student under this new supervisor. At the time I was in my mid-twenties, still very shy, but I was coming out of my shell some. He made a couple of comments to me that made me feel like I didn't fit in. One comment related to the fact that I wore Levi jeans instead of Wranglers. He told me I would never be a range conservationist because

I wore the wrong brand of jeans. He also commented on my shoes, a pair of white Nike tennis shoes that were my standard work shoes. Again, he commented that I wore the wrong shoes and would never be a range conservationist. He made the comments in front of a crew of fire fighters. I was humiliated, but I didn't respond and just kept doing my job.

It never occurred to Beth to leave the Forest Service. She said, "To be honest, I never considered leaving the agency because of these experiences. I am stubborn and things like this only provoke me into working harder to prove these kinds of people wrong. I have a strong belief that what goes around comes around. I'm now a district ranger and thanks to these experiences, am stronger and much more vigilant over my female employees than I ever would have been had I not had these things happen to me."

She was not aware, either, of any other women who left the Forest Service because of male resentment or discrimination, and said, "As a matter of fact, there are many women with longtime careers with the Forest Service that have similar stories to mine. They are strong women who have done great things during their careers. Sometimes, these types of experiences challenge you to be better and stronger. That's what I've seen from the women I've worked with." That she "just kept doing her job," and doing it well, is evident in the steady progress Beth has made in her career.

Figure 153. Beth on Stash in Bronco Grazing Allotment.

As a seasonal range technician Beth worked with grazing permitees in allotment administration. She conducted range analysis, consisting of Parker Three-step Clusters, paced transects, estimating range type, condition and trend, production and utilization surveys and allotment inspections. She drafted allotment analysis maps from aerial photos. Her duties included construction and maintenance of range and wildlife improvements, such as fencing and water source developments, trail construction and planning and implementing district transportation plans.

Wildlife management issues are inherent in range management. Range and wildlife resource managers work closely together, and sometimes the duties of both are combined in one job. Both range and wildlife work, particularly in the desert, often require travel into remote, sometimes hostile locations, on foot, horseback or by floating rivers, and camping out, often for days. Beth's duties included caring for Forest Service horses and mules. She was comfortable around the animals, and a skilled rider. She became skilled at loading stock and tack into horse trailers and driving them over treacherous roads. She learned to navigate rivers by kayak.

During one of her seasons as a range technician Beth supervised the district wildlife survey crew, conducted wildlife surveys, censused Arizona hedgehog cactus, northern goshawk and Mexican spotted owl. She performed wildlife habitat analysis and timber stand delineations. She did cone crop monitoring on Mount Graham in New Mexico, related to habitat for the federally Endangered Mount Graham red squirrel. She did similar work as a range conservationist, plus watershed analysis and wildlife reintroduction.

Her first permanent position was as a range conservationist on the Cave Creek Ranger District of the Tonto National Forest in Cave Creek, Arizona. There she did grazing allotment management planning and implementation, including permit administration, monitoring, supervision of work crews and range improvements (Figure 153). She provided project support for a complex district program, leading or supporting

Figure 154. Beth and rancher flagging stream in Lime Creek Grazing Allotment.

interdisciplinary teams, preparing biological evaluations and conducting archaeological surveys. She was a member of a forest-wide team assigned to develop sound methodologies for measuring impacts on stream banks and riparian vegetation in Sonoran desert stream systems (Figure 154). On the side, she was a Level II law enforcement officer, a fire fighter, and a support dispatcher during wildfires.

After five years as a range conservationist, Beth became the assistant zone wildlife biologist for Mesa and Cave Creek Ranger District of the Tonto, moving her work location to Mesa, Arizona (Figure 155). Program management included budgeting, project support and implementation. She was an interdisciplinary team member for district and forest-level projects, and worked on a Forest Plan amendment for prescribed fire and fire use. During one year she was the project team leader and wildlife biologist for the forest-wide Tonto Focus (NEPA) Team.

In 1998, she promoted to the Wildlife, Fisheries and Rare Plants Staff on the Mogollon Rim Ranger District, on the Coconino National Forest in Happy Jack, Arizona. She had similar program management duties for a large wildlife program, and a lot more public interaction. She was the project team leader for an environmental impact statement for a range management plan. She was district liaison with Diablo Trust Collaborative Group, a collaboration of over 250 members of ranchers, environmentalists and private individuals, researchers and scientists. She was a member of the team developing the Anderson Mesa Pronghorn Antelope Herd Implementation Plan, a facilitated process involving eight agencies, organizations and environmental groups, and the Forest Service representative to a team developing a management plan for pronghorn antelope in Arizona. She also served as a member of a forest-wide team assigned to respond to litigation regarding management indicator species, wetlands and livestock grazing, and as acting district ranger.

Beth's next position was as wildlife staff for the Pawnee National Grassland, on the Arapaho-Roosevelt National Forest and Pawnee National Grassland in Greeley, Colorado. She managed a complex wildlife program, and worked with outside agencies and organizations on wildlife-related issues (Figure 156). She was the large project team leader for complicated and contentious district-wide projects. She was the research project coordinator for a large-scale research program that gets national and international interest. She also coordinated a

Figure 155. Beth on Hindu in Mesa Parade.

Figure 156. Beth at a golden eagle nest on Pawnee National Grassland.

large volunteer program with members of various prominent organizations. She obtained for partnership funding through the Audubon Society's Lois Webster Fund and Colorado Division of Wildlife to collect, summarize and analyze Dr. Fritz Knopf's extensive work with mountain plovers.

Next Beth promoted to Forest Wildlife Biologist on the Apache-Sitgreaves National Forest in Springerville, Arizona. She provided leadership, expertise and advice related to the wildlife program forest wide. She led, provided program oversight and expertise and administered the wildlife program for five ranger districts and the supervisor's office. She was the wildlife team leader and lead wildlife biologist for the Wallow Fire Burned Area Emergency Rehabilitation (BAER) team, supervising a team of several wildlife biologists from the Forest Service, Arizona Game and Fish Department (AZGFD), and the USFWS. She developed and managed the forest's wildlife and fisheries budget. She completed an assessment and report for known occupied or historically occupied threatened, endangered and candidate terrestrial species within the 538,000 acre Wallow Fire area, and was the lead biologist on emergency consultation for the Wallow Fire, leading a team of four biologists in completing biological assessments and initiating emergency consultation. She implemented monitoring of the post-Wallow Fire. She secured partnership funding through agreements with multiple agencies to accomplish forest-wide monitoring requirements and large-scale projects.

In 2015, Beth promoted to district ranger on Sacramento Ranger District, Lincoln National Forest in Cloudcroft, New Mexico. She described her typical workweek:

> On Monday I attended a field trip on the Sacramento Allotment with a group of New Mexico state representatives, office of the governor, Congressman Pearce's staff, local permitees, the deputy Regional Forester for the Southwest Region, the acting Forest Supervisor for the Lincoln National Forest, the state engineer, state agriculture department, and a host of others to discuss endangered species management and fencing that was constructed to protect habitat for the newly listed New Mexico meadow jumping mouse and its critical habitat. It was a highly charged and contentious meeting that lasted all day.
>
> Tuesday I participated in our weekly leadership team video meeting first thing in the morning, then a line officers call after that. Immediately after that I was in a conference call with the National Science Foundation and New Mexico State University regarding future management of the Sunspot National Observatory. After that I caught up on emails, then met with one of my employees to discuss bad behavior during the previous week. At some point during the day, one of my employees informed me his wife had been getting strange phone calls on his home phone threatening that all government employees were going to die. At the same time, I received a phone call that there was an oil spill on one of the district roads crossing the Agua Chiquita creek, where designated critical habitat for the New Mexico meadow jumping mouse occurs. The oil spill was contained in a pothole directly adjacent to the creek, but it was raining and had high potential to spill into the creek. The

oil spill was a result of our contractor's masticator engine that blew apart on its way to our newly awarded $750,000 partnership project with New Mexico Game and Fish Department. I spent much of the rest of my time on the phone with dispatch coordinating activities on the oil spill, and with forest service law enforcement and the county sheriff's office coordinating statements on the threatening phone calls.

Wednesday I went to the field with my recreation staff, communications tech, and forest engineer to meet with the county emergency services director to discuss putting an additional command repeater in a building near a historic lookout tower. I was able to get back to the office just in time to catch up on phone calls and emails and turn around to go out to the oil spill that had happened the day before.

Thursday morning I met with Congressman Pearce and his staff, NM state representative, the deputy regional forester, acting forest supervisor and the same permitees in the Rio Penasco to come to some agreement on actions we would take to reduce the impacts of the electric fence on the livestock grazing operation. The state representative told the deputy regional forester that it was time for a change in leadership (meaning me). This was the first time since I've been here that there has been a request to remove me by a high ranking official, but probably won't be the last time. The deputy regional forester declined. The afternoon was spent writing up an agreement documenting the actions we agreed to and timelines for completion. I sat down with my biologist and range staffs and updated them on the actions we needed to take and made assignments for the next week. My hotshot superintendent also came in to talk and update me after a two-week assignment.

Today, I was in the office early to speak with an employee and my deputy district ranger about communications issues in the timber shop. We spoke for approximately one hour. Once they left, I had multiple employees coming in asking for signatures on cruise designs, administrative restoration of compensatory time, updating me on the digital sign we are going to purchase and place outside the district to keep people informed, and a whole host of other things. My 40-hour week was completed early today, so I'm leaving early to go home for a long holiday weekend.

Beth bought a beautiful house in La Luz, New Mexico, a "dream home" that provides her the solace she seeks after a demanding week. It is on three acres sitting on top of a hill with 360-degree views of the mountains above and the Tularosa Basin and White Sands below. She is divorced, and lives there with three dogs and eight cats.

Her marriage dissolved due to the demands on her time related to her job. There are challenges living in the remote locations where Forest Service jobs are located. It was difficult for Beth's husband to find work in the small communities where they worked and lived. She and her husband had to make irreconcilable choices about whose career to chase. She chose not to have children partly because she did not think she would be able to do her job well and raise children well. She said, "I think everyone's challenges are a little different, and to be honest they can result from the choices you make. In my experience, the biggest challenge I've had is nothing to do with the work itself, but with the choices you make if you stick with this career. I've moved so many times with the Forest Service that it's hard to count, at least ten times over my nearly thirty-year career. It's difficult to promote in place and it's difficult to get well-rounded, in-depth experience if you don't move. There's a personal sacrifice that comes with that. Besides losing my marriage, I have left more good friends behind than I care to think about [though she has many good friends, most of them current or retired Forest Service, and she is still in touch with many of her high school and college friends]. I've left communities that I loved and started building friendships and community relationships over and over. There are financial sacrifices with all these moves. Buying and selling homes frequently, moving without regard to housing market values, costs associated with buying and selling homes in less than strong markets in remote locations, all add up."

Still, for Beth the reward has outweighed the sacrifice. She said, "I'm not the same person who started working for the Forest Service in 1984. I think I'm better. I think I'm stronger and smarter. I think I'm much more outspoken and courageous.

Working for the Forest Service has given me a level of confidence that is hard to describe. I believe it has built character and a certain charisma. I have a real sense of pride and satisfaction in the things I've accomplished and in how far I've come. It has tested me physically, mentally and emotionally, and I've grown through all of the years and experiences. I'm still growing. I'm still being challenged."

Janet Krivacek, District Ranger

Janet (Figure 157) was born in and spent her early years around the agricultural and farming community of Dixon, Illinois, population 12,000. A short distance from town, her family had a small hobby farm, which her mother managed, along with a large garden and multiple fruit trees. Her father's veterinary clinic was located there too. She said, "We had a few of everything, partly due to clients giving my dad animals." She was a 4-H member, and showed sheep, horses and garden goods that she raised.

Her family's values, illustrated by her dad's incredibly long hours in his veterinary practice and her mother managing the clinic and the home, was to work hard, but enjoy life and the company of other people, as well as be kind to all of earth's creatures. It was also important to her mother that they go to church.

Family recreation usually centered on water or the woods. Summer time highlights were to go fishing or boating with dad. And with mom it was mushroom hunting, bird watching or "tramping in the woods," as her mom would say.

Janet had both boy and girl friends, and being raised in an agricultural community, the girls that she had as friends were interested in similar outdoors activities.

During high school, due to her father's new job, her family moved to Two Rivers, Wisconsin, along the shores of Lake Michigan. She played a variety of team sports — softball league and high school track and swim teams and synchronized swim team. Point Beach State Forest was directly

Figure 157. Janet Krivacek (left) early in her career — Juneau RD waiting for helicopter to fly to stand exams.

across the road from her family's home and she enjoyed walking in the forest and along the beach with friends and family. She said, "The sense of being out and discovering what had changed from the last time I had passed a particular way because of the change of seasons or weather made traversing the same routes enjoyable. In college and beyond, exploring new country was fascinating."

Janet's favorite school subject was "probably math," with the natural sciences being a part of her life with exploring the outdoors and living in a rural environment. But she also liked art, especially watercolors, and being creative. During high school she submitted several watercolor paintings to a contest hosted by the *Milwaukee Journal*.

Originally, Janet thought she wanted to be a dental hygienist, as "everyone in my family had some type of medical degree." But her mom suggested she check out the natural resource school at the University of Wisconsin at Stevens Point and "the rest is history." It made more sense since she loved being outside and going on adventures with her mom to the woods. She felt it was a further extension of discovery and learning about the outdoors, plants and animals and how things work together and function.

She said, "I think what I loved about being a district ranger was I got to think and have discussions with employees and the public about all the resources, how they are connected and then how best to manage them for the long run. I also feel incredibly honored to be hired as a district ranger when I think about the history of the Forest Service and the integral part the ranger played. On one of the two districts I managed, I was the first woman ranger. On the other, I was the second. I feel lucky to be part of that whole history."

While Janet did not deliberately choose a non-traditional career, she says the summer jobs that she had, such as corn detassling and mowing grass, were done by both males and females, so she did not think much of it. She said, "My supportive parents encouraged me to try new things and explore activities that I had interest in. They were both outdoor people and they fostered an environment that I could accomplish whatever I wanted to do. I also had all sisters so I didn't see a separation of duties and activities that maybe occurs when there are brothers in a family also. If I had an interest in an activity that my dad was involved in or that he needed some help with, I went along or helped. I really didn't have a conscious feeling that I was going into a field that was non-traditional."

In college she was definitely a minority in the natural resource classes. She said, "I'd guesstimate less than ten percent of my class was female and only a fraction of those women went on to work in their field for more than a couple of years."

Early in her Forest Service career she was confronted with challenges being in a field that had been previously dominated by males. It was her perception that it was because she was a woman. "I thought as long as I could do the work and do it well, there should be no difference." She said that what made it okay was that in the majority of the instances she had great supervisors that realized what was happening, acknowledged something was not right and did something about the negativity. She said, "I didn't get why some people thought I should not be there." Though she never felt she had been assigned as a token, she did say, "It's a possibility I got a preference in hiring due to a manager wanting to diversify a staff."

Janet has experienced male resentment and outright discrimination. One time early in her career her supervisor wanted her and her female timber co-worker to get experience in timber cruising, and one of the males on the separate cruising crew was supposed to help teach them. The next day the man came up to Janet and within inches of her face told her loudly that they were not a part of the cruising crew and that they did not need to learn these techniques. In that instance, the supervisor set the man straight.

When Janet was a silviculture crew leader in the wilds of Alaska, a male crewmember did not like that she was directing the crew. She had a heated debate with him about who was in charge. The man actually pushed Janet down. He later apologized, saying that his religion made him believe that women should not be in the lead.

A positive challenge early in her career was that she had started her permanent career as a technician with the goal of becoming a certified silviculturist. She said, "When I found out that I needed to be in the professional forester 460 series and then be nominated to go thru a certification process, I realized I had some steps to go through. True to my upbringing, I set goals and with determination, I got converted to a forester 460 series and later gained support from supervisors and line officers to complete the process to become a certified silviculturist."

Janet's career spanned over thirty-four years, with the first nine years as a technician and the remaining years as a professional forester or district ranger. This included five regions and nine different national forests or regional offices. When she was a ranger, she and her husband lived apart during the week for four years due to their respective Forest Service job locations. They were able to reunite when she transferred to her last position in Missoula, Montana.

Janet acknowledged that while the entry of women into the Forest Service changed the agency, working for the Forest Service has also changed her. She feels incredibly lucky to have worked for the Forest Service when she did, seeing remote parts of the country, living the lifestyle, meeting extraordinary people and having a lot of fun working in the outdoors. She has had some incredible adventures, which have further affirmed her need to be self-reliant and confident and her desire for adventure.

She said, "I can't imagine having worked my whole career in an office whether it was in the Forest Service or some other occupation. It is unfortunate that as you advance in your Forest Service career that you spend more time in the office. I wonder if the old time rangers or employees felt the same way. You see these iconic pictures of the early 1900s ranger on his horse looking over the countryside. I wish that had been me!"

~

Traute Parrie, District Ranger

My request to interview Traute (Figure 158) came when she was twenty-nine years into "this unconventional career of mine." She wondered whether her story was worth sharing with anyone other than her family. She credits each of her family for parts of her success.

Traute was raised about 100 miles from Laramie, Wyoming, in Saratoga, where a certain amount of self-sufficiency was required. She is the oldest of three children, but was raised with lots of stepsiblings. There were seven kids in a three-bedroom house.

Her memories of teen years in small town Wyoming include starting the day by hauling water to the horses, or chopping ice and feeding hay and grain, which entailed a daily

Figure 158. Traute Parrie.

walk past a waterfall. This was the 70s when it was still common for winter temperatures to hover around twenty degrees below zero every morning, sometimes thirty below, and in her memory, forty below at least once. That was her quiet time. Then she returned to a house of seven siblings sharing a single bathroom. They walked to school, across railroad tracks, the Platte River, through downtown, past the student union with the ping-pong table and jukebox.

Traute and her mother shared a horse hobby, which Traute said was not far removed from their heritage, since her mother was raised on "the ranch." That was Springfield Ranch, homesteaded in 1902, by Herbert William Small, in southeast Wyoming, and recognized 100 years later as a Centennial Ranch. Management of the ranch went to her mother's older brother, but her mother stayed fairly close, at least while Traute and her siblings were kids, and they returned often to help, or to play. Traute reflected, "My memories are endless — lemonade breaks during haying, sleeping on the porch, the arts and crafts house that great-grandpa built with its innovative electric light fixtures, two-wheel drive pick-ups that finally gave way to the new four-wheel drives, cooling the watermelon in the irrigation ditch, driving the cattle to summer pasture, riding double on Bucky or Brownie, always early mornings, with the farm report on the radio, home-made ice cream! Canning beans. Grandma's ironing machine, and the cream separator. Especially playing double-solitaire with grandpa. But we were always the city cousins (from the big burg of Laramie, population 24,000!)."

Traute speculated that her lifestyle choice could have gone either way. Her sister Shirley ended up in New York City at age sixteen, and moved to Alexandria, Egypt for a few years. But something on the ranch "must have taken root" for Traute — the whole land stewardship idea, and she stayed in the West, working outside.

In college she had looked first at medicine, since three of her four parents were in the medical profession. But her dad was discouraged about the future of health care, even back then. (She said, "Thank you Dad.") She reached a critical juncture when she turned down attending the School of Architecture at Yale, choosing the University of Oregon instead. Her family was devastated at first, since to them Yale represented opportunity. Traute thinks her family eventually concluded that she would not have survived at Yale.

Traute came to a different conclusion, believing she was stubborn enough to survive. She was driven to independence by watching her mother, who had been divorced, and then widowed. Traute did not want to be dependent on anyone, and said, "I have a keen need for independence. I valued freedom and needed to go my own way. All the times I've heard 'you should not' (not go on fire, not have this job that belongs to a man, not take the kids to training with you) has left me cautious about seeking approval."

After rejecting a career in medicine, Traute looked towards the work of her dad's father, who built houses for a living, and his father, who built barns. She already loved working with her hands, and had won awards for the furniture she had built. Continuing in a creative line of work like architecture seemed like just the right fit. She was also strongly influenced by her stepfather Cotter's pursuit of a long-term energy solution. She said her continuing interest in energy resources and reducing our reliance on energy resources, is a legacy gifted from Cotter. It was Cotter's cancer that brought her back to Wyoming, from Oregon, and she finished her education with a degree in architectural engineering at the University of Wyoming.

As a child of the 70s, Traute could not help but be influenced by the feminist movement. She definitely did not want to be a nurse, but she did want to contribute to a better world. She was not the nurturing type, to the point of being combative. Having had to cook for six siblings as a child was a chore. Teaching involved too much conformity. She said she sees the consequences now of society turning a whole generation of women off teaching. She described herself as a nerd — glasses, good grades and good at math. She said that girls were unkind, but her brothers and their friends welcomed her. Consequently, a career in a male dominated field like engineering, held enormous appeal.

Traute's eldest child, Lindsay, was born ten days after her graduation from college in 1982. She worked in construction until the birth of her son Zach two years later. At that point, she read a story in the local paper about the retirement of an engineering technician from the local ranger district of the Forest Service, and she contacted that office about the possibility of a job. It was easier to get on with the Forest Service as an engineer, considered a shortage category at that time, than as a forester. Thus began Traute's land management career.

Traute recalled that she was thrown into the thick of it right away, unlike trainee positions she observed later. She learned by "on-the-job" training, with no field review of her projects by supervisors until final inspections. In the end, though, she said this was the best learning she could have had — following her designs from planning through implementation, and learning from mistakes.

After about four years, she faced another decision point. The forest was facing litigation from environmentalists for the first time, and created a brand new project planner position, called a NEPA coordinator — the first ever on the Medicine Bow National Forest. Traute had reservations, as it meant coming in from the field, and working much more behind the computer. She took the job, and was forever grateful for the exposure she gained working with other specialists such as the silviculturist, archeologist, biologist, ecologist and others. She especially enjoyed her time on interdisciplinary team field trips, watching how each of these specialists viewed the woods from their unique niches. Her four years in this position expanded her horizons.

She used her engineering roots to follow opportunities as a zone engineer in Laramie, Wyoming followed by a move to the Bitterroot National Forest in western Montana, where she served as the forest engineer. In 2004, she took a departure from the Forest Service to work for the BLM in her first assignment as a line officer. She was the associate field manager for the White River Resource Area in Meeker, Colorado, where she worked for the first time with a major energy development, as well as with wild horses. She returned to the Forest Service in 2007, taking the district ranger position on the Beartooth Ranger District of the Custer National Forest (now the Custer Gallatin National Forest) in Red Lodge, Montana (Figures 159 and 160). She held that job until her retirement.

Throughout her career, Traute spent considerable time qualifying for and serving in various militia positions in support of fire fighting. She went out as a crew member or squad boss on a Type 2 hand crew from 1985 to 1996, then she got on a Type 1 incident management team as a resource unit leader (RSEL) in the planning section of the team. She moved to Region1, and was on a Type 2 team as a RSEL from 2001 to 2004. She also had other assignments over the years, as a single resource such as dozer boss trainee on the Bitterroot, and RSEL on hurricanes Katrina and Marilyn. She kept her basic fire fighter qualifications until the end of her career, so she could

Line Officers

Figure 159. Traute (center) at Hogan Creek looking at a grazing allotment on Beartooth Ranger District.

Figure 160. Traute (left) delivering a talk "when the moon is full" along the Meeteetse Trail, Beartooth Ranger District, Wyoming.

help with prescribed burning, which she said, "At least once required digging line to keep it from getting away from us — and just so the troops could see me suffer through the arduous pack test, always good for a little entertainment." Once she got her ranger job, her fire duties revolved around the agency administrator role and prescribed burning.

Fire assignments were twenty-one days long in her early days, and she would be on multiple assignments during a season. She said, "Fighting fire was a huge source of crew cohesion and bonding that carried over to other aspects of the 'day job' and my personal life. There was a holistic element, not to mention the glue that held so many of us together. That fire experience still defines at least a part of me."

Traute said that whatever confidence she may have with people has been hard won, but that confidence with hard work was a given. After years of early morning chores, she was not intimidated by her first fire assignment. She said, "A Pulaski is not that different from the axe handle."

She poignantly described the reward and hardship of fire fighting:

> Memories of night shift: moon rises to sun rises, sounds of falling Douglas fir snags in the northern California forests, adapting hose lays as belay ropes, deuce and a half rides, misty mornings in Oregon spike camps, turning my fire boots upside down so snakes wouldn't crawl in. Camaraderie! Days spent learning about different resources from each other as we constructed endless miles of line. The burn-over on a ridge in Idaho in 1992. Extended shifts without support — no food, no gear but what was on our backs.
>
> Fire fighting gave me still more confidence in my physical abilities, not to mention the opportunity to explore — and has led me to the most rewarding connections of my life. These are bonds built in hardship. The radio traffic from a nearby fire that turned out to be the Storm King fire with fourteen fatalities while I was on a Colorado ridge above De beque in July 1994, is seared into memory. The shared experience with my crewmates there will never be replicated, and those bonds have yet to be severed. On 9/11 (The 9/11)

when crews had to be pulled off the line because air support was grounded, along with all other aviation across the country, we shifted our focus from the fire to putting fire fighters in touch with their scattered loved ones. A few weeks later I served in the Forest Service honor guard at the funerals of a fellow fire fighter who had pushed and helped me to pass the pack test earlier in the year. Much earlier in my career, when it was my turn to finally go out on initial attack, a co-worker and I were first on scene on a fire that turned out to be a plane crash. After Hurricane Katrina, when I was anxious to put either my engineering or emergency response background to work, I spent a couple of intense weeks practicing frenzied adaptation and working side by side in the ghostly aftermath with the most interesting people in the world.

The common thread through all these events has been the immersion in the human experience — weariness, sorrow, discovery, joy, serenity and camaraderie. I remember the landscapes, and even the hardships, but *it's the people I've met* that made the biggest impression, fellow travelers in this unique walk of life. It keeps me humble, and gives me faith about the future of societies/community.

A trend that troubles Traute is the fact that there are fewer women in the wildland fire organization now. One crew she was assigned to in 1987, had more women than men, but these days she does not see even a healthy minority of women on the fire lines. She thinks it is similar in the field of engineering. She wonders whether our culture lost ground in valuing our scientists — and not just for women.

Traute lived in Wyoming for sixteen years, staying put so her kids could be raised with their extended family. She is divorced from their dad, and married Don Carroll in 2000. She and Don have four children between them. They raised their kids in the backcountry, "hopefully setting them on their own paths to confident and independent thinking and experiences." Their grown children all have chosen careers that combine personal passions with the idea of leaving things better than they found them. Traute said, "These children are my greatest accomplishment."

Traute has harbored a lifelong curiosity about other cultures, and had traveled some with family or friends (Figures 161 and 162). Her ease in the outdoors has kept her looking for ways to "combine my love of the outdoors with behaving responsibly with respect to community and as a land steward."

In recent years she was asked to apply her Forest Service experiences in new landscape conservation efforts in Africa. The Forest Service International Programs works overseas for what they can contribute and what they can gain. They link skills of field-based staff with partners overseas to address the world's most critical forestry issues and concerns. Traute's assignment was the logical next step in her lifelong path of learning and exploration. She had initially volunteered to work in the Boma Jonglei protected areas of South Sudan, before independence, but the team was never successful in getting travel visas at the same time, given what was going on politically, so she was then directed to work in western Zambia.

The project she worked on there was meant ultimately to support the Zambian government with wildlife habitat and forest management, particularly the issue of deforestation in the face of climate change. Traute had a lot to learn about social structures and governance including the role of the tribal chiefs, education, or the post-colonial culture of aid dependency in so many African countries, with some of its unintended consequences. There was the particularly relevant phenomenon of "visitor fatigue" experienced by local villages who hear so much talk from all the visitors but who often see no discernable results. Traute and her team exercised adaptability to try to provide something meaningful for their African counterparts, and came away with unending respect for their newly found friends.

Three weeks in Africa was such a brief exposure to a continent in such rapid transition, but for Traute it was a great moment of immersion into several aspects of this African country, ranging from simple, clean and welcoming villages, or the trade communities along the main commercial highway, to the capital city Lusaka with its cosmopolitan enclaves. That trip

Figure 161. Traute (standing) looking down toward Chola Tsho, Nepal.

Figure 162. Traute with Everest, Cholatse, and Taboche, Nepal.

was just the beginning of her African education, and she hopes to continue to find a way to stay engaged.

Traute recently retired from the Forest Service, and is proud of the legacy she leaves and is optimistic about her future. She summed it all up with, "I am the child of all my ancestors; the ranch ancestors, my carpenter grandfather, my uranium prospecting stepfather, my trail hiking father and step-mom, and my independent, willful and strong mother. You gave me the skills to survive. You modeled them, and expected them of me. You needed to have confidence that I would survive, but in my own fashion. You have my enduring gratitude."

∽

Ruth Wooding, Last Sula District Ranger

Ruth (Figure 163) grew up, through high school, in Moorestown, New Jersey, near the seashore outside Philadelphia. She spent lots of recreation time at the Pine Barrens, and her parents eventually moved there. Moorestown was a large town of about 15,000, a beautiful, old-fashioned, middle class town. Ruth was exposed to a mixed society, where she experienced no bigotry. By contrast there was extreme poverty and seventy percent minorities in nearby Camden, New Jersey and Philadelphia, Pennsylvania.

Ruth loved being outside, being in the woods and "Down the Shore" on the Atlantic seashore. She was a good student and loved science, which was her passion in high school. By seventh and eighth grade she knew she wanted to work in the forest after meeting a ranger in Acadia National Park.

Her father owned a corporation. He was an electrical and mechanical engineer, and an inventor. Ruth said, "My beloved father was brilliant, slightly eccentric, beyond passionate about his family and a proud capitalist." Ruth is the fourth of five girls in her close-knit family. She credits her values to her parents and like her "mummy" considers herself an independent.

Ruth's parents emigrated from England. Her father was mostly Irish, and her mummy British, both Catholic, and both

Line Officers

Figure 163. Ruth Wooding — Ranger Rue at her beloved Sula Ranger District.

deeply religious. In England, during World War II, they received wartime degrees in engineering and teaching, respectively.

Both of Ruth's parents have a profound love of music and took lead parts in musicals like *Pirates of the Penzance* in Britain. They emigrated in their late twenties. The "Wooding Family" sang at the Sunday morning Catholic Mass in Moorestown, Mummy playing organ. The children went to Catholic schools. The girls sang in five-part harmony with training from their parents. During a ski trip one year to Sugar Bush, Vermont they met the famous Von Trapp family in the Our Lady of the Snows Parish Church the Von Trapps had founded. Both families sang together for that midnight Christmas Eve Mass "taking the roof off" as Ruth recalled. All the girls in Ruth's family are musicians, singers and guitarists, and in their teen years played in bands and coffee houses in New Jersey. Ruth continued her coffee house playing later on the West Coast.

Her father was firm and her parents were "strict and unique and loving." Her dad would laugh and say, "When you girls turn eighteen either get out or pay rent!" They knew he really meant it, and they were all out on their own after eighteen.

They were expected to go to college, to be self-sufficient. Her parents sent her four sisters to Wales and England for college, because Ruth's father felt Americans were "parochial" and that the girls needed to experience other cultures. Ruth defied her father and went west for college and as a result she was the only Wooding girl who had to work her way through college, with Mummy slipping her money to get by. Her Dad ended up being proud of her. She had defied him because she "knew with all my heart I was meant to be a forester and work in American forests and therefore did it on my own."

Ruth's parents are avid outdoors advocates. Both did a lot of walking, and her mummy is a remarkable photographer of small microsite nature. Ruth sees her mummy as a true one in a million woman raising five girls, teaching first grade full time and, at ninety, volunteering for her church and the food bank, and still playing organ and keeping her home on a Maine lake as neat as a pin. Mummy managed a large household with skill. Both of Ruth's parents imparted a strong work ethic to their children.

Ruth had a pony, and was in the English riding circuit for many years on the East Coast, beginning in sixth grade. During grade school she worked after school and weekends for the horse farm to earn money to help pay for the pony and stables.

Ruth credits her career choice to her love of science and the outdoors, and to her introduction to the west during a family ski trip to Jackson Hole, Wyoming when she was thirteen (Figure 164).

Now she loves public lands, saying, "I'm as crazy about them as I am about my own children." She has a passion for having a part in stewardship and protection of the lands. She loves trees, considers herself the quintessential "tree hugger" and literally hugs trees regularly on her walks in the woods. She still has the same sense of awe and wonder about trees

Line Officers

Figure 164. Ruth (second from right) with dad and sisters, Jackson Hole, Wyoming.

that she had as a child. She still finds the woods as her comfort space, a place of solace and quietude, where all her stress unwinds. She makes a point of being a part of nature in her daily life: gardening with a subsistence garden, huge flower beds and walking in the woods weekly in order to stay connected with the land.

Ruth went west in 1977, at age seventeen the fall after graduating from high school, and worked full-time in Seattle, Washington while waiting to get Washington State residency. Her first job there was a night job on a production line making sandwiches. She was violently ill the first week there, working production belts that were unsanitary. She worked there for a month and "never ate a wedge sandwich again!" She found a friend there with a music industry connection. The friend worked at Norlin Music Inc., a company that owned Epiphone and Gibson Guitars. It was a high-end musical instrument inspection business taking in instruments from Japan, tuning or repairing them after their long ocean voyages, before selling them in America. In December 1978, Ruth met the manager of the Seattle Norlin Music Epiphone Branch and was hired for a job inspecting Epiphone Instruments. She befriended other musicians while there and was part of a small band that played open mike nights and coffee houses.

After she earned her state residency, she was accepted to the University of Washington, College of Forest Resources, to study forest management.

In 1980, to better fit her class schedule, she got a gardening job with the University of Washington botanic gardens, Washington Park Arboretum. She worked there as a Master Gardener and as the arboretum gatekeeper for the rest of her college years. She was thrilled to be given the Stone Cottage in the park to live in on-site. The Stone Cottage was built by the Work Progress Administration in the 1930s, and was a precious small cottage in the heart of the park. Ruth worked at the arboretum part-time during school, and full-time during summer managing the gates nightly and in the morning and at a second job at a nearby Seattle nursery. She graduated from UW College of Forest Resources, then one of the top ten forestry colleges in the country, in 1983.

In 1984, Ruth got a seasonal job on the Mount Baker Ranger District, part of the Mount Baker-Snoqualmie National Forest, in reforestation and silviculture. She did timber sale inventory (TSI) and layout for timber sales: layout of units, stand exams and cruising. She worked alone and sometimes felt terrified by getting turned around in the old growth forests of the Pacific Northwest (PNW). Many of the areas Ruth worked in were magnificent "cathedral" old growth forests on extremely steep slopes with monstrous Douglas firs, western red cedars and western hemlock with a thick mix of devils club, vine maple, alder and big leaf maple understories. The PNW woods are notorious for disorienting hikers who get lost in those woods, some of who were never found again. As terrifying as the deep woods could be at times during her often solo work, she loved the job. She said, "I could not believe they would pay me to be in the forest I loved so much." When fire season came she completed basic fire fighter training and served on Type 2 fire teams on huge "project fires" in other states. Fire work included being dropped into raging wildfire areas by helicopter and "spiked out" in wilderness areas as happened in the Frank Church River of No Return Wilderness Gospel Hump fire.

In 1987, she got her first permanent U.S. Forest Service job as the reforestation specialist for the Big Summit Ranger District of the Ochoco National Forest in Prineville, Oregon. It was a hard transition due to the job's low pay and higher income taxes in Oregon; she resigned in the winter of 1988, and went back to a seasonal job on the Mount Baker Ranger District, this time in engineering. Within three months she got a permanent job as the survey crew chief. Her duties included surveying, road location and running a chain saw to clear preliminary lines. Ruth had a certification for tree climbing at this time, climbing old growth white pine trees to collect blister rust resistant cones, using climbing spikes and rappel ropes to gather cones and rappel back down. Ruth was the only woman in her class to complete the course and get her climbing certificate.

This was a time period when Earth First was monkey wrenching in the woods and pad locking themselves to logging equipment, putting sand into expensive back hoes and yarders and spiking trees with iron spikes that went to the mill causing injury and mill shut downs. Earth First ripped out miles of preliminary line survey stakes and ribbons that Ruth had established as the survey crew chief, destroying weeks of work with a five-person team. They then pounded the ribbons way up on an old growth tree's bark after climbing the tree to put them there. Ruth received an award for an innovative design to outsmart the monkey wrenchers. She planted ten penny nails on point, with dummy stakes precisely four inches behind them, then when Earth First ripped out all the stakes the survey crew just went back in and relocated each survey point with a metal detector.

Between 1988 and 1991, Ruth took two advanced automated computer aided (AutoCAD) drafting classes at Skagit Valley Community College, and became one of three AutoCAD technicians for the engineering division of the Mt Baker Ranger District. In 1991, she leveraged that job into being the only engineering technician for North Cascades National Park Service Complex. She did small site design, road design, earthwork calculation and cartography maps for recreation sites. She designed a Hilfiker retaining wall for the approach for the 300-foot Skagit River Bridge leading to the Newhalem North Cascades National Park Visitor Center (Figure 165). Using her survey skills she measured and calculated earthwork removal for the bridge approach and used her survey leveling skills to help the contractor install it. Ruth used AutoCAD to design trails after locating them and surveying them in the woods. She was capable of being a one man survey crew (replacing a crew of four) by bringing traffic cones and a level rod and her buckers tape (an auto wind up steel tape measure on a roll in a metal case) to substitute for a crew, and then used that data to design one-foot AutoCAD contour topographic maps to use for small site design. Her survey and cartographic skills lent themselves to her becoming the accessibility coordinator for the park, designing facilities and access paths for persons with disabilities. She took sustainable design courses and designed the Hozomeen employee cabins at the top of Ross Lake in Hozomeen, Washington inside the park. The cabins were made of log and had solar panels — joined twin log cabins installed and functional at a remote site replacing older less comfortable and energy inefficient employee housing.

She created opportunities for the elderly and for those with strollers. She has a particular passion for providing universal opportunities for disabled access, since one of her sons has a heart condition, an invisible disability that creates stamina issues. Level access paths allowed everyone to enjoy the forest. Ruth incorporated varying degrees of slope challenges into the

Figure 165. Ruth on portion of Skagit River managed by the Forest Service as part of the Mount Baker-Snoqualmie National Forest.

access trails, working with "the remarkable maintenance crew of the park where everyone caught the passion to make opportunities in the woods for the disabled." For more advanced chair users, she worked with her dear friend Kevin who had been paralyzed in a car accident, who tested the trails with his wheelchair.

In 2000, Ruth was promoted to the private lands and easement administrator job role for the Sawtooth National Recreation Area in central Idaho. She and her husband bought a log cabin on eleven acres in Challis, Idaho and she became the only Sawtooth employee ever in its thirty-five year history to commute from Challis to Stanley, Idaho, a 130-mile round trip up the Salmon River Scenic corridor the locals called the Gauntlet. She was infamous in the community for this commute but it was an easy choice for Ruth since there was no high school in

Stanley for her boys and she preferred to commute four hours daily than have her children do it on a bus. Challis was a tiny ranching town, not too particular to federal employees, but they always treated Ruth and her family with love and respect because she was part of their community.

She brought ninety conservation easements, with a value of $44 million dollars, up to compliance through annual inspection of each easement. Many had not been inspected for more than a decade. When an easement is not inspected annually and has unauthorized developments it is impossible to go back and have the developments removed. Easements are issued "in perpetuity" and annual monitoring allows the government to ensure they are protecting the scenic values for which they were established. For this work Ruth received an award for her dedicated service to the landowners and the Sawtooth NRA by the Sawtooth Society for her "Dedicated Service, and Gracious Spirit (Figure 166)."

Figure 166. Ruth receiving Sawtooth award for collaboration.

During her time at the Sawtooth Ruth was nominated as a member of the twenty-person Upper Salmon Basin Model Watershed Project technical team, a multiagency organization out of Salmon, Idaho. With this group she worked collaboratively with other agencies towards the acquisition of more than a half million dollars in grants for ranchers in the Sawtooth to reconnect the main stem of the Salmon with its smaller tributaries by removing barriers. The group also obtained grant money to build scenic log worm fences to keep cattle out of out of the Salmon River's 750-mile long salmon migration route, to protect the fragile salmon redds. It is the longest salmon migration run in the country.

Ruth received another promotion at the Sawtooth, due to accretion of additional duties. The accretion of duties is an unplanned situation that develops over a couple of years, and if it is found that the incumbent has acquired new higher-graded duties during that period and that the agency needs the employee to continue doing them, then the employee can be promoted in place without having to compete for the position.

Ruth was a member at large on the Sawtooth National Forest leadership team for two years, and during the last five months of the private lands job, she also served as acting deputy area ranger for the Sawtooth NRA.

Ruth's next job beginning October 2007, was the district ranger position on the Sula Ranger District on the Bitterroot National Forest in Sula, Montana. That was the pinnacle of Ruth's career. She said, "I never dreamed of making it that far in the Forest Service, because as a young punk you did not even look in the district ranger's office." Her responsibilities included administration of a 260,000-acre district and responsibility for an inholding community of about 800 people who lived up a dead end twenty-mile highway in the wildland urban interface. Her responsibilities also included Lost Trail Powder Mountain Ski Area and its permit and the forest's only recreation residences. Sula was the third largest ranger district of the four districts that made up the Bitterroot National Forest. The Bitterroot is in one of the smallest national forests in the Northern Region of just 1,500,000 acres. Sula was a unique historic ranger district, and had the largest compound on the Bitterroot National Forest, with four residences and an old ranger station, most of them historic, a historic barn, a big bunkhouse, a fire cache and fire warehouse and large pastures. Ruth's pride and joy was making the ranger station neat and tidy with re-paved parking lots, freshly painted signs, and newly painted and spruced up buildings. She fought to get a maintenance person for the district to install a new sprinkler system and ensure the lawns were watered. She oversaw a pollinator garden of native plants in front of the station, and purchased some of the plants herself.

The historic East Fork Guard Station is on the district and Ruth fought hard for funds to have the picnic shelter built there in the style of the CCC era buildings.

Ruth's activities on the Sula district varied. Team management took about fifty percent of her time. She gave awards for good performance, helped several employees with grade increases and also held people accountable.

Figure 167. Ruth's photo of Sula Peak lookout tower area.

Line Officers

Figure 168. Famous photo taken during wildfires of 2000, Bitterroot National Forest, Sula Ranger District.

The forest averages more than 150 wildfire starts per year due to lightning strikes, and the Sula district had a large initial attack fire organization (Figure 167). In 2000, there had been catastrophic fires on the Bitterroot that burned one-third of the forest. Half of the Sula Ranger District was burned (Figure 168).

There were lots of dead trees and falling snags, which resulted in the death of an employee before Ruth arrived as the Ranger. There were a lot of long-term employees on the district, who Ruth described as a very talented team. They were devastated by this incident, and were still dealing with it when Ruth arrived on the job.

At the time of Ruth's tenure the District was the Range Center of Excellence working with private ranches that had rights to graze, pasture and stock cattle on National Forest System lands. Sula also had a large recreation program, including the Anaconda-Pintler Wilderness Area and the Sapphire Wilderness Study Area. Ruth also oversaw permit administration for the Lost Trail Powder Mountain Ski Area, one of the premier powder hotspots in the country. There were flocks of Bighorn sheep and an enormous elk migration route on the district averaging 7,000 elk.

Ruth implemented the Middle East Fork Restoration Plan within the 6.5 years she was there, cleaning out the interface of dead and dying trees near the community, resulting from beetle kill. No one thought the project would happen because of economically tough times for the timber industry, but with compromise, adjustment and hard work it was all completed, providing jobs and safety to the community.

Day-to-day, Ruth worked with contracting and loggers. She built a shelf stock of projects with advanced planning, ready when funding opportunities arose. She kept her eye on the ball and kept programs moving. She made a point of getting out on the ground and into the woods. She was results-driven and always looking for projects.

Programs on the District were unique and challenging, given the Lewis and Clark history in the area and tribal influence. The Lewis and Clark main encampment on the Bitterroot was near the ranger station and the area had significant cultural resources.

When Ruth came to the Sula district, she was working for Dave Bull, "the best supervisor on the planet." She said he preserved the integrity of the forest's four districts. He was gender-blind and a mentor. His faith in her was important to her. She had received the highest performance ratings and many awards throughout her combined Park Service and Forest Service career to that date, and she continued to thrive under Dave.

Dave's retirement in 2009 coincided with major downsizing, culminating in the consolidation of the Sula district, both sad events for Ruth. Sula employees who did not retire or resign were reassigned to the supervisor's office. In 2013, Ruth left her beloved position for a planning job in the regional office. It was devastating after such a glorious career but she felt grateful to still have work. She holds the bittersweet distinction of being the last ranger to ever work on the 100-plus-year-old Sula Ranger District.

Ruth chose a non-traditional career deliberately. She was always a tomboy, and has a comfort level working with men.

Her mother was a teacher, and her sisters are doctors, social workers and bankers. She is proud of having entered a male-dominated field and serving the public. She said, "My sisters think it's cool I was a ranger. I look at myself through their eyes when I get frustrated. They call me "Ranger Rue."

Ruth has a couple of times had "impossible supervisors" who did not share her work ethic, took credit for work they did not do and who hurt and bullied employees. In the course of her career she says, "I still find it wonderful, I am enthusiastic and I still love working for the agency."

When she transferred to the regional office she moved her family to a gorgeous log cabin on a four-acre farm in Stevensville, Montana across from wild open spaces: the Lee Metcalf Wildlife Refuge, a Buffalo Ranch and the Bitterroot Selway Wilderness in the distance (Figure 169). The family is self-sufficient, growing their own food, and hunting for meat. Ruth likes raising her children rurally, protecting them from outside influences. She loves having family dinners "with candles and flowers, home cooked gourmet food and the comfort of a conversation across the table."

She works hard and plays even harder. She spends her weekends outdoors, walking and visiting refuges and the national

Figure 169. Ruth's home in Stevensville.

forests she loves. She is grateful for having had the hard work of being in the woods throughout her career, doing some fairly death-defying activities, in deep mountainous forests.

Ruth's extraordinary experiences include being on huge project forest fires, being active and physical, "climbing steep mountain sides, being in the deepest old growth forests and driving death-defying gravel roads in jeeps!" She once smashed her elbow when she was thrown from a mule on a trail ride. She described her "most mind-blowing, earth-shaking experience," which occurred at the age of thirty when she was working as an engineering technician for the Park Service:

> I was doing FERC (Federal Energy Regulatory Commission) relicensing inspections of four dams on the Skagit River that provide the bulk of the hydropower in the Pacific Northwest along with the Columbia River. My boss told me later he didn't think I would make it past the first day of the four-day inspection. I visited four dams and was blown away to learn they were all hollow! I went into the Newhalem, Gorge Dam, Diablo and Ross Lake Dams and their power stations. The Gorge was the smallest of the four on the mighty Skagit River. I love the Skagit River and often went steelhead fishing there. At Newhalem and the Gorge Dams the FERC inspector I was with literally climbed over the side of these 200 to 300 foot vertical drop dams so I had to follow along on the vertical ladder. I walked over the edge of the Diablo Dam onto the sharply angled spillway — thinking I was nuts to do so — finding a dynamite bore, a rock core six inches long removed to place dynamite, which I still have today. Ross Dam took my breath away. It was 500 feet tall, with internal ladders (vertical half stairs on a 45 degree angle dropping 500 feet straight down into a black hole) leading into wet cubbyholes and walkways with piezometers measuring pressure far into the bowels of the dam. Unbelievably there was a chair at the bottom cross walk where it is said the ghost of Ross Dam lives; a worker who fell to his death during concrete pours. His body was never recovered. As I climbed up and down the ladders of all four dams, my reaction was 'you've got to be kidding me!"

Mary Erickson, Forest Supervisor

Mary (Figure 170) knew mobility from the start of her life. She was born in Boise, Idaho, where her father was a doctor at a VA hospital. She attended kindergarten in Fairbanks, Alaska, where she experienced the 1964 earthquake. Next she attended first grade in Salt Lake City, Utah. Then came a summer in Michigan, followed by second and third grades in upstate New York, in a small town near Albany. Fourth grade through high school were spent in Neenah, Wisconsin. Mary's mom was from the area, so they had a lot of relatives there. They lived near Oshkosh, with lots of agricultural influence.

There were four children in the family — one older sister, one older brother, and a younger brother. Mary was the most outgoing and sociable. She was the ringleader who organized events and brought friends home. Her siblings were more reserved. Mary liked the outdoors, always had a dog and spent lots of time outside with the dog. She liked getting dirty and playing in water, "catching critters." She was always plotting activities.

Figure 170. Mary Erickson.

They were tight-knit, spending lots of family time together, playing games outside. They had a playhouse. Their indoor activities included such games as Risk and Ping-Pong. They did not camp as a family, but took driving trips, once from Wisconsin to Alaska. Mary's father was more of a debater and talker than an outdoorsman. Her mother had lots of brothers, and more practical, outdoor skills. She liked the outdoors and took the kids outside for activities including cross-country skiing.

Mary's parents were conservative Catholics, and Mary and her siblings attended church and Catholic schools. Mary liked school and most subjects, especially math and reading. She felt she was not as good at art. Always industrious, Mary graduated early from high school by doubling up on classes in her junior year of high school. She planned to move to Oregon for college, but her parents would not let her leave until she was eighteen. She attended a semester at the University of Wisconsin, and then at eighteen, she moved to Corvallis and enrolled in Oregon State University's forestry program. Out-of-state tuition was expensive, so she dropped out long enough to get her residency, and then got her Bachelor of Science and Master of Science forestry degrees there.

Mary was attracted to forestry because she was interested in doing something outdoors and active. She also admitted that she was being rebellious, since her father wanted her to be a doctor.

She had been influenced by attending the Trees for Tomorrow Camp in northern Wisconsin, which was suggested by a high school counselor. She had a close cousin who had moved to Oregon and liked it, so she researched forestry schools in Oregon.

Mary loves the Forest Service for its opportunities to live and work in great places. She believes in public lands management. She said, "I like the people in the Forest Service and the issues the agency deals with; not so much the controversy and being on the firing line, but the ability to influence long-term public success and protect lands for the future."

Field-going positions came early in Mary's career, first while she was waiting to obtain Oregon residency. She worked then for Oregon State Forestry as a forestry trainee, as a dispatcher and fire lookout. She also worked for the university system in summers, doing inventory for the Oregon State MacDonald Forest.

In 1984, between her bachelor's and graduate degrees, she was hired by the Forest Service as a co-op education student, in the position of analyst and economist on the forest planning team. It was an indoor job, which was the downside, but she liked the interaction with forest professionals, the exposure to experienced people and to the public process side of the business. She got early exposure to all the public interests the Forest Service deals with. She stepped into leadership roles with the forest planning team as opportunities presented themselves. After only two years in her permanent job, she led a briefing of the Chief and under-secretary of the Forest Service in Washington D.C.

In early 1988, Mary was asked to fill an acting district ranger position on the Chemult Ranger District of the Winema National Forest. She got the position when it was filled permanently in late 1988. She loved it there and determined not to ever leave, and did stay there for eleven years. She did a fair amount of travel as part of the position, serving on regional committees and Washington D.C. assignments. As district ranger she oversaw about eighty-five permanent and fifty to sixty seasonal employees.

In 1999, the region consolidated the Winema and Fremont National Forests, and Mary was asked to step into an acting deputy forest supervisor position for the two forests. The Chemult district ranger job was filled behind her, and after two years as the acting deputy forest supervisor for the Winema-Fremont forests, Mary got a forest supervisor job on the Fishlake National Forest in Richfield, Utah. She served there for six and one-half years before going to Bozeman, Montana as forest supervisor in December 2007. She was not in Bozeman long before another consolidation between the Gallatin and Custer

National Forests resulted in her becoming the forest supervisor for both those forests. The Custer was headquartered in Billings, Montana, and extends into North Dakota. The consolidated land base of over 3.1 million acres is now dubbed the Custer Gallatin National Forest, and Mary manages it from the Supervisor's Office in Bozeman.

Mary's day-to-day now includes a lot of travel, meetings, conference calls and videoconferences and lots of management. The two forests do not have perfect alignment of their issues; the complexity of issues and the geographical scale of the forests stretch Mary's time. On the Fishlake National Forest, with its closer districts, she could be more intimately involved in details; it was easier for the leadership team to function as a team.

She spends lots of computer time to keep up and communicate. She feels less connected to the ground, but still makes the effort to get out and make decisions based on what she sees. She strives to visit each district in the field twice per year. On weekends she and her family get out to see more of what is happening on the forests.

Mary does not feel that she deliberately chose a non-traditional career, or even the Forest Service — mostly she wanted to work outdoors. She loved the outdoors and the west; she really didn't know much about the Forest Service. A woman she met along the way, Nancy Graybill, a silviculturist, pointed her in the direction of the Forest Service.

When she worked for the State of Oregon early in her career, the workforce had been male-dominated. By the time she went to work for the Forest Service in 1984, there was a mix of male and female in planning, and there was a good mix on the ranger district at Chemult.

Mary met her husband on the Chemult Ranger District and they married in 1992. Later in her career, when she started moving around, it was a challenge to manage dual careers. Both worked for the same agency. They had to consider how and when they could move and both have satisfying careers, and what they were willing to do as a family. Mary stayed in place longer than she would have otherwise. The Forest Service was very supportive, but it was still a challenge. There was one job in Region 5 that Mary was offered at the same time that she was about to adopt her oldest daughter, and the region was unable to accommodate Mary's personal needs, so she declined the job.

She loves the land that the National Forest System manages, and living in places of natural beauty. She likes getting out on the land. She said, "I like the challenge of the national forest management arena, the tension of different expectations from different people. I love working with people for better outcomes." When she first started her career she loved work life and making a difference, and still does, but once she adopted her two daughters, her family gained the edge of importance. She has no desire to move for another promotion; family considerations, providing grounding for her two daughters is more important. Adopting the girls in China — the travel to China, meeting other adoptive families and bonding with them and the whole raising kids process, is at the top of her list of extraordinary personal experiences. She said, "Raising good, happy, healthy kids who enjoy a successful life is most important to me."

She told of a couple of her extraordinary work experiences. During her second year on the Gallatin she went on a three-day horseback ride into the Absaroka-Beartooth Wilderness. Riding into the high country where there were no people and experiencing the amazing country was memorable. Mary considers anytime that she gets into the backcountry, away from other people, under her own power, to be a special experience, all her own. She said, "I still enjoy the wilderness vicariously."

Another time in the Taylor Fork with fisheries employees she saw her first grizzly bear sow with cubs. The bear stood up and looked at them. Mary described the bear as being about fifty feet away, though it was probably more like 100 yards away.

Mary said she has no regrets about having chosen her line of work. She is honored to be in the Yellowstone area, "though the bureaucracy can be a pain." She has learned the necessity of being open-minded. She loves the ability to choose great places to live, and "no big city!" Though she now works mostly in an office, she loves that the work is focused on the outdoors.

She still loves being able to make a difference, both in land management, and in "having the influence to facilitate making others' careers fulfilling, making decisions on who to hire and support." She said, "I would put employees in the Forest Service against any other profession in respect to being good people, great employees who are good land stewards, and respectful of all interests."

~

Patty Grantham, Forest Supervisor

Patty's father spent a career in the U.S. Coast Guard that kept the family moving between the east and west coasts of the United States as he changed assignments. Patty's earliest memories come from the early 1960s, when the family was living in southern California. They lived in the suburbs, surrounded by farmland and a little aerospace industry. She remembers riding her bike everywhere with a "pack of friends." In the late 60s, at the age of nine, the family moved to New York City. At that time, the Coast Guard had a significant base on Governor's Island, which was a little piece (two square miles) of real estate in New York harbor just off the tip of Manhattan. They spent four years there, getting the best of the city experience coupled with the security and quiet lifestyle of a military base. From the wilds of New York City, they next moved to the very different wilds of Juneau, Alaska, around 1973. While it was the capitol of Alaska, Juneau was a sleepy little town back then. Patty said, "I was a teenager with a love of the outdoors. What could be better?" The family ended up in the Seattle area, where Patty finished high school and attended the University of Washington.

Patty (Figure 171) characterized herself as a "tomboy through and through while growing up, although I did have plenty of Barbies to balance out my GI Joes." She does not

Figure 171. Patty Grantham with grand twins Danny and Nora in 2014.

remember having any girl friends in grade school, and felt more affinity with boys than girls through her thirteenth or fourteenth birthday. She said she was not necessarily bored with girls' activities, she just found "the stuff boys were doing much more interesting." She did the typical activities of a kid of that era — played outside mostly, rode bikes. As she got older, she spent more time hiking and being outdoors in a more exploratory way. She played basketball in junior high, high school and college. She enjoyed the physical and mental challenge of the game, the competition, and learning to work as a team with others.

She learned to sew and bake from her mother at an early age, and to work with her hands in carpentry from her dad. Those activities have been lifelong pleasures.

She has three sisters. Her parents had the girls in two sets eight years apart, Patty being the third one born but the elder of the second set. She said, "About the time I was developing into an interesting person (if I do say so myself), the older sisters were off to college/getting married, so my younger sister and I had the luxury of being in a smaller family unit as we moved into late childhood/early teenager hood." She has fond memories of her family camping, both with all of her siblings and then with just her and her younger sister. They hiked and enjoyed the outdoors. Patty said, "In some ways, this was surprising as my father spent most of his formative years in Chicago and my mother in coastal Virginia where being outside meant working on farms. I'm not sure where they got the idea that they should take us camping." She and all of her sisters were active in Girl Scouts, with their mom as a leader and their dad as a supporting father.

Patty liked most school subjects. She liked reading and music class as much as she liked science and math. She related a funny story about being in about fifth grade:

> It was at about that time that we started taking vocational tests. I remember taking two or three of them in the period of a year or two. These were to help kids figure out where their interests lay and what kind of jobs would be a good fit. My tests kept coming back with strong matches in architecture and engineering. The strongest match, though, was in agriculture. I could never figure that one out. I liked gardening, but I didn't want to be a farmer. It took me until I was late in high school to realize that agriculture was the closest thing those tests had to forestry. I guess the tests were pretty accurate after all.

Patty did not experience the sense of being a minority in any of her classes. She was in the honors track in high school, along with lots of other girls. She attended the University of Washington, College of Forest Resources (now School of the Environment) from 1978 to 1982, and there were about twenty-five to thirty percent women in her forestry classes. She said, "It wasn't too lopsided. I guess I was really lucky to have chosen the University of Washington."

Patty's eldest sister was in college when the family was living in New York City. Her sister got a summer job working at the Statue of Liberty where she met and ended up marrying one of the park rangers. He was transferred to a small national park in upstate New York shortly after they married. Patty and her younger sister would visit them there. Patty said, "I quickly determined that my brother-in-law had the greatest job in the world. He got to work in this really cool place, be outdoors, explore the woods, help people — it was all so wonderful. So, that is what I set out to be initially. Once I started forestry school, I became more interested in the forest management side of forestry than the interpretive side." She does not think she deliberately turned to the non-traditional. She said, "I think I knew what I liked and wanted to do it."

Patty said that every one of her thirty-five plus years with the Forest Service has been great. From 1980–1982, she worked seasonally as a forester for the Caribou National Forest on the Soda Springs Ranger District in southeastern Idaho. Most of her time was spent in timber management and silviculture, though she also supported recreation and fire. There were about ten permanent employees there, and they all did everything.

Patty worked in progressively higher graded positions in silviculture, fire, land management planning, lands and minerals, recreation, wilderness and trails. She worked from 1982–1986, as a forester on the Clearwater National Forest, on the Powell Ranger District in Idaho. She primarily worked in silviculture in TSI/reforestation, with time in timber sale planning, recreation and fire. From late 1985 to mid-1986, she also had a long-term temporary assignment on the Gallatin National Forest in land management planning. She helped compile the first interagency Greater Yellowstone Plan.

From 1986–1990, she was a land adjustment forester and promoted to a lands and minerals staff specialist in the supervisor's office on the Mount Baker-Snoqualmie National Forest in Washington. She worked in the land adjustment program (purchases and exchanges) initially, and then moved to the cooperative roads program, but also did support in special uses, minerals and small hydroelectric project permitting.

Next she left the supervisor's office to work from 1990–1992, as a district staff officer on the Mount Baker Ranger District. She supervised a large recreation, wilderness, trails, wild and scenic rivers, lands and minerals program. During the summer and fall of 1991, she also served as the acting district ranger.

She then transferred to Alaska as a district ranger on the Petersburg Ranger District of the Tongass National Forest, from 1992–2007. The ranger district covered 1.7 million acres and had large programs in timber management, roads, fisheries and recreation. She spent about fifteen months in 2000–2001, on a temporary assignment as the forest recreation, lands, minerals, heritage and wilderness staff officer.

She returned to the lower forty-eight in 2007, as the deputy forest supervisor and ultimately the forest supervisor on the Klamath National Forest in northern California.

Of her career, she said, "It's been amazing, often challenging, and always interesting. I've stayed with the Forest Service because it's been a great place to work. I've always been given as much challenge as I've wanted to take on. I feel like I've been respected and listened to. I've gotten to work with a bunch of people who liked to work hard (like me) and were deeply committed to serving this notion of public lands. I've gotten to do some really neat things and work in amazing places. It's been more than I could have ever dreamed."

Patty said that one thing she loved about her being a forest supervisor was her ability to make the job easier for employees working in the field. She said, "I think that most of what I do can be pretty invisible to forest employees, but making sure that the local and regional political and community relationships are strong brings tremendous benefits. I also have to make sure that the forest, strategically, is moving in the right direction. My current job is challenging; I seldom know what will walk in the door day-to-day. The other very important thing I love about my job is the ability to influence the next generation of leaders in the agency. Supporting people young in their careers to grow and learn and take on progressively greater responsibilities in leadership is very satisfying. Somebody did it for me, and I hope to be giving that back."

As far as challenges go, Patty does not think anything she has faced is unique. She said, "I have experienced disagreement with male colleagues, even felt great disdain for some of them (and the feeling was mutual). I don't think these experiences were born of male resentment or discrimination, though. I just think it was basic human dynamics." She remembered one instance when things "maybe got strange." She was being considered to attend training that she needed for her job, but ultimately did not get the training. She said, "Later I heard that the wife of a male colleague who would have gone to this training with me, approached my ranger with the concern that her husband would be travelling with and attending this training with a single woman. According to the old adage about believing 'none of what you hear,' I don't know if this really happened. But, to this day thirty-plus years later, I remember this story, mostly because it hurt me."

She started her career in southeastern Idaho as a single woman in a conservative community. She was the first or second natural resource professional woman to work at that ranger district. Older gentlemen came into the office wanting a wood permit and were told that they needed to "talk to Patty about that." The gentlemen would say that they would "wait till the men got back" and leave rather than work with Patty. When the gentlemen returned, they asked for her (male) boss who directed them back to Patty. She realized later that that was probably not an easy thing for her boss, a long time resident, to do in a small town that had a certain view regarding jobs women should be doing. She said, "I think that I was an anomaly in this office and small town and I think that some people who I worked with probably didn't know what to think/do with me early on, but we all got over that. The other very helpful thing is that there were a few young professionals on the district and on a neighboring district, just starting out like me, and we hung out together. The support I've received over the years from the agency, my supervisors and my coworkers has always been phenomenal."

In the mid 1980s, Patty was one of the first (if not the first) female fire crew bosses in Region 1. She recalled going to a morning fire briefing and being told that she could not have a shift plan (these days, known as an Incident Action Plan or IAP) because her crew boss already had one. She had lots of experiences along those lines in fire back then. She does not think people were trying to be sexist or nasty — it was just that seeing a female in a fire leadership position was outside their scope of experience.

When she went to Alaska to be a district ranger, she was a single parent with a four-year-old son. Juggling work and travel commitments with being the only parent was tricky. She said that her son was a good sport about it. Later, after she remarried and her daughter was born, she and her husband both had active travel commitments. She remembered many times when they "transferred" the baby at various airports in southeast Alaska due to one parent needing to start a trip while the other was still on the way home.

Patty's life goal has always been to care for her family and support her children to be successful. In her life as a whole, her most extraordinary experience has been the birth and rearing of her two children. She said, "Having kids is the best thing I've ever done."

Her greatest work goal has always been to leave a place better for her work there. She said, "Few people can change the world overnight, but most people can set improvements in motion that will make things better, if not today, then next week or next year. I harken back to Edward Abbey's famous quote about not burning yourself out, about making sure that the driven-side of you is balanced with the enjoyment-side of you. Being in forestry is about being in the long game."

Patty said that she has had so many incredible experiences at work that it is hard to pick just a few, but getting to work in Alaska, in truly wild and remote areas with incredible fish and wildlife resources has got to be one of them. She recalled being on a field review once, and discussing with other agency personnel the placement of some key wildlife reserves within a project, when suddenly a wolf pack from across the drainage began to howl. She said, "It was as if they knew we were talking about them."

She got to live and work at Powell Ranger Station, a historic and remote place at the headwaters of the Clearwater River. She said, "Living on that remote Forest Service compound connects a person to the history of the agency, to how it was done in the old days. It was a great experience, one that is getting more rare all the time."

Another extraordinary experience was having had the honor to meet every Chief of the Forest Service since the beginning of her career.

She has had the chance to go to Grey Towers a number of times, including once to attend a meeting of a small group with Teddy Roosevelt IV. Being at Grey Towers brought the agency's history to life, and Patty wishes every employee could go there.

The last really great experience she wanted to mention was getting to participate in the agency's Senior Leader Program. She said, "I can't begin to explain how much that experience meant to me (and still means to me today). I got to build career-lasting relationships with my forty fellow participants/peers, a group of incredible people that I've come to rely on for advice and counsel. The program also instilled great self-awareness in me and really gave me the right skills to be successful as a leader (and a person). It is the greatest gift the agency ever gave me."

Patty acknowledged that working for the Forest Service had changed her, and she summed up her feelings about her career with the following:

> Being a part of the long history and culture of the agency has imprinted on me. Most of the change that has happened in me is one of those things where I already had a tendency toward doing that thing. I grew up in a military family. Serving America was a big deal. I grew up in a volunteering

family. Serving others was a big deal. I loved the outdoors and am completely enamored by the systems of the natural world. As I've grown older in my career, I am deeply gratified by seeing younger colleagues who I may have helped gain success and grow to be leaders. As I've dealt with tragedy at work, it has changed how I view the gift and preciousness of life. Some of these things I may have experienced working for any employer, but many of them not.

I marvel at my dumb luck at landing on forestry as a career and with the Forest Service as my employer. As I get older, I realize that, with a few pretty subtle differences in "the route taken," who knows where I would have ended up? The younger people coming into the service and their energy and commitment inspire me. They aren't going to do it the same as me (thank goodness!) and at the end of their careers the agency will be the better for having them and the changes that they will make. I am grateful.

Leslie Weldon, Deputy Chief for the National Forest System

Leslie Weldon's father was in the Air Force, which afforded her family the opportunity to live in several states and overseas. She was born in 1961, in Pullman, Washington, where the family lived briefly, and then she spent several elementary school years at an Air Force base in Japan. She was raised primarily in the principle family home in Oxon Hill, Maryland, a large metropolitan area outside Washington D.C. in Prince Georges County. There were six kids in the family, three boys and three girls.

Leslie (Figure 172) liked "hanging with the family, close to home, without a lot of other friends." She liked the outdoors, and was involved in Girl Scouts in Japan and in the United States through eighth or ninth grade. In middle school through high school her favorite subjects were English, biology and all science except for chemistry.

Her family members were not extreme outdoor enthusiasts. They did some camping a couple times per year. They did

Figure 172. Leslie Weldon and Smokey.

a lot of crabbing and clamming in Japan and on Chesapeake Bay in Maryland. Leslie did some hiking in Girl Scouts.

Leslie was attracted to her career in the Forest Service by her interest in the outdoors. In the 1970s, she was in late elementary to middle school, when there was a lot of environmental awareness and pollution prevention going on in society. When the Endangered Species Act passed, it sparked her interest in natural resource management. She did not deliberately choose a non-traditional path. Her parents had always told her that she could do anything she wanted to, and she simply followed her interest.

At the age of fifteen she fell into an opportunity through an upperclassman to join the Youth Conservation Corp (YCC), and spent the first of two summers in a residential program in Maryland. During her second summer, she was a youth leader. The job was at Camp Rocky Knob for the National Park Service in the Blue Ridge Parkway. Her crew built trails, split rail fences, maintained campgrounds and some cultural sites. They attended conservation education classes every Friday. As a result of that experience, Leslie chose Virginia Tech as the college she wanted to attend.

Leslie entered the agency in 1984, midway through the first wave of women hired by the Forest Service for other than clerical work. She said, "Three attributes made me different than the traditional white male forester typically employed by the Forest Service; I was female, African American and a biologist." During her second year in college two African American men from the Forest Service in the Pacific Northwest recruited Leslie for a job. The Gifford Pinchot National Forest and Job Corps were looking for minorities.

She experienced mostly subtle discrimination due to her differences. An example is a district ranger who told her he did not think she was cut out to be a co-op student, which she thought was strange coming from a leader. She did not take it to heart, and harbors no resentment over it. Overall, her experiences have been very supportive to her career pursuits.

Leslie started her distinguished career with the Forest Service as a seasonal on a reforestation crew, monitoring seedlings, fighting forest fires and surveying spotted owls. During college Marsha Carney of the Forest Service got Leslie into the co-op education program as a fisheries biologist, as there were more opportunities in fisheries than in wildlife at that time and place. After receiving her degree she was hired as a fisheries biologist for three districts on the Mount Baker-Snoqualmie National Forest in Washington where she worked restoring habitat for salmon and steelhead.

Between 1987 and 1991, Leslie served as Assistant National Fisheries Program Manager and as a staff biologist for the Northern Region. She was assistant ranger and district ranger on the Stevensville Ranger District of the Bitterroot National Forest from 1992–1996, where she led extensive local public involvement efforts to implement collaborative, ecosystem-based management of forestlands. This included launching the long-term Bitterroot Ecosystem Management Research Project in partnership with the Forest Service Inter-mountain Joint Fire Sciences Lab and University of Montana.

From mid-1996 through 1998, Leslie served with the Northeastern Area, State and Private Forestry, as Forest Service Liaison to the U.S. Army Environmental Center at Aberdeen Proving Ground, Maryland. She pioneered an interagency partnership for technical assistance in natural and cultural resource management on U.S. Army bases and other military installations stateside and overseas.

From 1998 through 2000, she was Executive Policy Assistant to Forest Service Chief Mike Dombeck. She assisted in managing operational priorities for the chief across programs for National Forest Systems, Legislative Affairs, Research and Development, State and Private Forestry, International Programs, Business Operations, Civil Rights and Financial Management. In November 1990, the Forest Service held a National Diversity Conference. Leslie had a strong role on the planning committee for the conference. She found it incredible that the Forest Service would hold such an event, and said that being a part of it was a highlight of her career.

From June 2000 through June 2007, she served as the forest supervisor for the Deschutes National Forest in Bend, Oregon. Leslie led management of natural resources, business operations and customer service for 1.6 million acres of the Deschutes National Forest in Bend, Oregon. Programs there included premiere year-round outdoor recreation, watershed and aquatic restoration, forest health restoration and fire management on complex fire-adapted ecosystems. Programs also included commercial forest products, heritage resources, transportation and facilities management, geology and minerals, fish and wildlife and extensive outreach and partnership programs. Leslie worked closely with people and organizations interested in the stewardship of the forest for various uses including strong interagency programs with the Bureau of Land Management.

Leslie was External Affairs Officer in the Office of the Chief from June 2007 to October 2009. She was responsible for several national programs including Legislative Affairs, Office of Communications, Media and the National Partnership Office.

From 2009–2011, Leslie was the regional forester for the northern region of the Forest Service. She provided oversight for management of twenty-eight million acres of national forests and grasslands across five states as well as State and Private Forestry programs in northern Idaho, Montana and North Dakota. On any given day, Leslie had significant conversations about issues raised by the public that needed her engagement. She provided guidance to twelve forest supervisors and to regional office directors about their programs and activities. She provided advice and counsel to individuals. She worked in close coordination with the regional forest leadership team (forest supervisors and directors) on strategic management and operational issues. She engaged with the national headquarters office on all significant issues. She had meetings with partners — groups representing the environment, industry and different uses of national forest lands.

In November 2011, Leslie Weldon was named Deputy Chief for National Forest Systems with the Forest Service. She is the lead executive responsible for policy, oversight and direction for the natural resource programs for managing all national forests and grasslands so they best demonstrate sustainable multiple-use management, using an ecological approach, to provide benefits to citizens.

Leslie has had a rich career. She has participated on numerous national and international technical and policy review activities. She has had the opportunity to travel with international programs. Her work has involved identifying shared goals and working closely with partners from academia, conservation and environmental groups, community groups, local government and natural resource related businesses. She is committed to workforce diversity, leadership development and civil rights.

A personal challenge for Leslie has been balancing her family life with work. She and her husband both have careers in the natural resource field. They have adult twin boys, so she has been a full-time working mom for her entire career. Because of the career opportunities that have come her way, she is her family's main provider. Balancing her need to be available to her family with work is challenging, but she said she has not faced workplace barriers that have made it harder.

While she never has developed a "greatest goal," she said her passion is "to be true to my core values. I take my leadership role seriously, and am careful not to abuse it She said, "I want people to be nice to each other." Reflecting on her career, Leslie stated:

> My chosen profession resonates with my desire for complexity, challenge and opportunity posed by sustainable natural resource management. I enjoy being able to provide leadership to the agency to accomplish sustainability in a complex environment. I like the interaction with citizens in the roles they play in managing resources, and I feel the agency can only be successful through that interaction. I like that no two days are ever alike; I am constantly exploring and learning. My personal and professional relationships that developed through my career enrich my life.

Abigail (Gail) Kimbell, Chief Emeritus, U.S. Forest Service

Starting life as an east coast girl, Gail (Figure 173) was born in Boston, Massachusetts and finished high school in St. Albans, Vermont. She was the second of eight children so she spent a lot of her childhood helping with younger siblings.

Active and curious, she enjoyed outdoor activities such as bike riding, swimming, canoeing, skating, hiking and fishing. Her family hiked together all through the White Mountains in New Hampshire and spent time on lakes fishing, boating and swimming.

Gail also always loved math and science. She related how the combination of that passion and her attraction to the outdoors drew her to her profession:

The ability to meld being out of doors with math and science was definitely an attractor. I grew up hiking on the White Mountain National Forest and very much enjoyed the people I met (fire lookouts, foresters and recreation guards). And I still love forests. Forests are increasingly important to us in the United States and around the world as sources of water supply and scrubbers for our air. They provide wildlife and human habitats. They provide wood — an amazingly beautiful and versatile material. The importance of quality forest management only increases with the growth in population and the estrangement from the natural world.

Gail earned a Bachelor of Science degree in forest management from the University of Vermont and embarked on her most productive career, starting as a seasonal in 1973. She worked as a forester for the BLM and the Forest Service, in

Figure 173. Abigail (Gail) Kimbell.

Oregon and Alaska. She went back to school for a Master of Forestry degree from Oregon State University in Forest Engineering. She was a district ranger twice, a forest supervisor three times, associate deputy chief, regional forester and then retired as Chief of the Forest Service. Of the scores of varied experiences spanning her career, she still counts among the most extraordinary an incident that occurred during her early field-going days:

> As a presale forester working out of Kodiak, AK, I was responsible for marking out timber harvest units and their associated roads systems. There was a crew of four of us camped out in remote locations for ten or more days at a time and often worked in the pouring rain day after day. After one exceptionally long day in the woods, we returned to camp, too tired to cook dinner. While we watched the red salmon swimming in dense schools along the shoreline just feet below where we sat, a brown bear came towards us along the beach. It was an extraordinary moment as we all looked at one another. Yes, the bear left us.

As her call to higher leadership evolved, Gail struggled with that responsibility versus her desire to stay closer to the ground at lower levels. She said she never set her sights on being chief. She did not seek out Washington Office assignments or details; in fact, she avoided going to Washington for twenty-eight years. Fortunately for all of us, she has "a high need to work for people I respect and to do work that is meaningful."

Others recognized her potential and urged her on. When Dale Bosworth became chief, he started recruiting field people into Washington. Gail had, on occasion, expressed some opinions on national policy and Tom Thompson, deputy regional forester in Region 2 at the time, told her it was time to "put my money where my mouth is/was and that I needed to take the detail that was being offered to me as associate deputy chief in National Forest Systems. And I did when I learned that Tom was to become deputy chief for National Forest Systems." Still, she said her heart has always been on the ground and though she gave it what she could as acting, she returned to her forest supervisor job in Colorado much wiser than when she left. She shared what she had learned in her Washington D.C. assignment:

> The folks in the Washington Office work very hard to buffer the field from some very crazy stuff and work in a wild environment trying to get good solid policy through the channel. It wasn't until I worked there that I appreciated the skills and talents there and the sacrifice that so many people were making. I know now that I would have been a far more effective forest supervisor had I gone to Washington first. It's important that as public land managers we appreciate how public policy is made and how a decision made in the field can affect the whole agency.
>
> Communications are so critical. That said, my own record doesn't always shine here.

In the end, with encouragement from friends, Gail did apply for and got the associate deputy chief position she had acted in.

The rest literally is history. She moved to Missoula, Montana as Regional Forester for the Northern Region and felt she was just hitting her stride. She said, "The moment when I was asked if I would permit my name to be submitted as part of a list to the Secretary of Agriculture for consideration as Dale Bosworth's successor, I was stunned. I was more than stunned. I could hardly breathe. Who would want that job?"

The job included a lot of actions that were unpopular with the field, including the move of human resources to Albuquerque. The agency's budget was being gutted by the cost of fire suppression but the agency's overseers wanted some kind of risk free fire suppression effort and everything else. Gail described her internal struggle by saying, "With climate change, our forests are subject to insects and fire like we've never seen before and yet there was only marginal public license to do much about it. Communities around the country had raised a flag of concern and warning about water quality and many about water quantity. Who would want the job of chief forester of the United States? On one hand, I clearly did not. On the other, on the chance I might be able to make a difference, how could I say I did not want to be considered?" As chief, she stated that to see successful collaborative efforts towards stewardship of public resources really excited her. She cited that her greatest goal during that time was to remain healthy and to be an asset to today's managers rather than a critic or a pest. She further reflected, "In retrospect, I am very glad I did assent to include my name. It was quite an experience and I hope I did make a difference in the focus on the issues for the change of administration. But it would have been so very nice to stay in Montana and deal with the issues from there, at the regional forester level."

Still, another of Gail's most extraordinary life experiences sprung from her role as Chief of the Forest Service. Serving in that capacity in 2009, she attended a Committee on Forestry (COFO) meeting in Rome, Italy — a United Nations meeting of heads of forestry from the world's nations. She was elected chair of the meeting that included discussions on many critical worldwide issues. Her first act was to recognize the gentleman from Afghanistan. She marveled, "Until then, I hadn't even wondered about the forestry resources of many of these countries but have since learned." Gail's track record weaves a rich tapestry as we see how she availed herself of the opportunities to continually broaden her horizons during her journey through her unconventional career.

Gail did not deliberately turn to a non-traditional career. She was influenced at thirteen or fourteen when she was part of a group listening to a talk by a forester at the Saco River Ranger Station on the Kancamagus Highway on the White Mountain National Forest. He was talking about how trees grew and how you could measure that. It was the first inkling Gail had that you could combine being out-of-doors with math and science.

She was always encouraged by her parents to pursue her interests. She said, "The absence of women in forestry seemed to me more of an artifact of the generations. Still, it was definitely a novelty to have women starting in field positions both with the Forest Service and the BLM. The novelty thing worked both for and against us."

When she started her career, there were very few women. She said, "My own experience was simply that, my own. For the women who chose to have families and careers, their challenges were multiplied." As examples, she cited working with a woman on the Willamette National Forest who was an exemplary forester and an exemplary mom, and another who was a mom and a contracting officer in Region 9 who had worked in multiple regions. Both made difficult career and personal choices that she was never faced with but greatly respects. She said both served the public extremely well.

When Gail started as a seasonal most of the men she worked for were her Dad's age and were World War II veterans. She said, "The addition of Viet Nam war veterans, war protesters/long hairs and women really rocked the world for a lot of those guys." Gail's early supervisors were at a loss what to do with women in their workforce. Do you send them to fires? Where do you camp them out? Do you give them real work to do where they might get dirty or get hurt? Gail said, "Yes, there were some challenges — certainly to be expected. I was lucky enough to find people who would assign me the hard work and to allow me to sink or swim. I was even luckier to find some people who would talk me through some of it." She said that the surprise for her was in people of her own generation discriminating against women simply because they were women. Of that she said, "It goes on today and I still find that surprising. But then, there is no law against being a jerk and there are enough of them out there to go around."

Gail is retired now and summing up her career, she said, "Sure there were moments when I would have chucked it all for a clean workspace in an air-conditioned or heated or dry office. Or for a job that allowed me to work alone and not have to go a single interdisciplinary team meeting again. But I wouldn't trade my career in public forest management for anything."

∼

Lauren Turner, District Ranger

I conclude this series of profiles with my own story. I am grateful to have shared the stage with these extraordinary women, to be a part of our history when we helped move society and the Forest Service forward by manifesting our love of the land through our work.

When I was four years old, my father moved our family to Yosemite National Park, nestled among the Sierra Nevada Mountains in California, for a seasonal job with the concessionaire there. He fell in love with the place, and managed to stay on and earn enough of a living to raise my four siblings and me in the wonderland of Yosemite. That first summer we lived near the river in a two-room canvas tent with a plank floor. When winter came we moved to a small trailer. As my father advanced with the company, our housing improved.

The surroundings could not have been better. As children, my friends and I played outdoors within the magnificent granite walls that define Yosemite Valley. We swam and innertubed in the brisk waters of the Merced River. We climbed trees and mountains, hiked, biked and rode horseback. We slept outside in the front yard in summer, and skied and ice-skated in winter. My family did a lot of tent camping. My dad took me fishing and taught me to tie fishing flies, and how to clean the catch.

Yosemite was small but offered many activities. I went to Junior Rangers, Bible school and swim lessons. I was in Brownies and then Girl Scouts. I enjoyed park ranger talks and an occasional orchestra on the lawn at the famous Ahwahnee Hotel, and stately formal teas there. I loved visiting the Indian museum and the library.

I was in one of the earliest kindergarten classes at the elementary school, and attended there through eighth grade. There were thirteen kids in my graduating class, and about ninety in the entire school. The school still stands, teaching even fewer today.

High school was in Mariposa, California, an hour and a half bus ride. I liked most subjects but did not really like going to classes. I had always been a shy kid, and did not adjust well to meeting new people in high school. I found ways to miss a lot of school, but still did well enough to earn scholarships for college.

I was in high school during the 1960s, a time of social change, much of which I was barely aware. I was, however, aware of the hippie movement, and attracted to it, the experimentation and "free love" aspects. And there were plenty of hippies who found Yosemite and who had an influence on me.

I met my first serious boyfriend, a couple years my senior, in Yosemite during my senior year in high school, and my life took a dramatic turn as I found myself a young bride and mother. My marriage lasted for about two years, half of which my husband spent as a draftee in the Viet Nam war. I lived with his parents while he was away, and attended community college classes part-time while there.

After my divorce I returned with my two-year-old son, Brian, to Yosemite for a couple of years. There I met and fell in love with my second and current husband, John. Miraculously, he fell in love with not only me but also with Brian, and we have been a family ever since. We settled for a while in Sonora, California in 1975, where we both attended community college long enough for me to gain clerical skills (there was a forestry technology program at that school I expressed interest in; I was in a government job training program and was told that the forestry technician tract was only open to males), and for him to receive a job offer that took us to Modesto, California. I hated that city environment, and when John and I visited his brother who lived in West Yellowstone, Montana, I convinced him to move there. It was there I drifted into my career with the Forest Service after about a year of waiting tables. I was hired as a seasonal information receptionist, which led ultimately to the career I never dreamed of.

We moved back to California after a couple of years. My transfer to the Lake Tahoe Basin was to be the first of a series of moves we made during my Forest Service career. Next we moved back to Sonora, where we planned to stay to give Brian some stability during his high school years.

A lot of my career felt like it happened by accident, but I think that growing up in Yosemite, with the influence of the Park Service, steered me toward work in natural resources. Many of the kids I grew up with ended up in careers with the Park Service or Forest Service. I did not seek a non-traditional career, but evolved into it. I was always attracted to more non-traditional roles, but they didn't seem open to me. My family was pretty traditional, my father clearly being the "head of the household." My mom worked, though, to help stretch dollars. I don't think that people who live in national parks usually get rich, and a lot of the moms worked in part or full-time jobs. I also had a friend who had a single mother. So I had some non-traditional role models in that respect. But I was pretty grounded in the 50s culture where dad worked and mom stayed home and took care of the house and kids. I even had fantasies about such a life for myself when I was young. John's family had a similar dynamic to mine. So my career came with a whole lot of changes that my family and I struggled through.

Mid-career, through what I consider remarkable serendipity, I found my way back to college, qualified as a wildlife biologist and started working in the field. I had about reached the limit for promotions in the clerical arena, and was struggling with how I was going to finish my education to qualify for something else when the consent decree opened up new opportunities. I got a scholarship, complements of the decree that solved the financial part of that problem. I was surprised that there were not many other women jumping at the opportunity.

Then came other practical and psychological difficulties. It was a three-hour round trip commute to school every day. That and study time impacted the amount of time I had for John and for Brian, who was in high school. Deeper than that, and harder to talk about, was the inevitable role-reversal that played out over time as I became more educated and the primary wage earner for our family. We worked through that, for years beyond my graduation from college. A positive that came out of it was that Brian and John developed a deeper bond as John had more of the parenting role.

I juggled the commute, study time and family responsibilities. Because the government was footing the bill, I felt a self-imposed pressure to excel. I graduated Magna Cum Laude with a 3.7 GPA.

During summers the Forest Service gave me work as a biological technician. I tromped around in the woods, setting up survey routes and surveying for spotted owls, goshawks and furbearers. I monitored snags — dead trees planned to be left for wildlife during timber sales, and located aspen stands for improvement projects.

Besides learning the fundamentals of wildlife management duties, the transition to field work had physical challenges, notably hiking alone in sometimes treacherous terrain while learning to navigate with map and compass (getting lost a few times), and figuring out what to do when being dive-bombed by an angry goshawk during a survey. Driving on horrible, narrow rutted dirt roads, sometimes facing-off with huge logging trucks cliff-side was a little nerve-wracking too. Fortunately I had some great mentors to help me. My supervisor, Kathy Burnett was the greatest. She taught and encouraged me, and hired me back for three seasons, giving me a strong foundation. And it was that way each time I changed course and promoted to the next level. There is always something new to learn, and there was someone to help. Once I got into fieldwork, I never felt discriminated against for being a woman, other than maybe an occasional subtle stray remark. Most of my struggles seemed to be adjusting my personal perception of what my role should be.

I had some of my most extraordinary experiences during those early days. The best of those involved moments in nature when I came upon a scene or had a moment when I felt whole, at one with my surroundings. I'll never forget the surreal feeling the first time I went to the woods with a few other biologists to learn how to "mouse" owls. Owls have incredibly sensitive hearing, and as soon as a mouse was placed on a log, a spotted owl came swooping in on silent wings, picked off the mouse and continued in a continuous arc back up into the trees, prey in its talons, seemingly oblivious to our presence.

A difficult hike is also etched in my memory, scrambling up steep slopes and crawling through thick brush, the whitethorn scratching my skin like cat claws. The reward was worth it. I rested on a huge juniper log, probably at least forty-eight inches in diameter that had fallen fortuitously at the top of a slope of smooth granite that spread out below it. I was high enough to look out over the top of the forest below. The sky was crystal clear blue, the sun just warm enough. My attention was drawn to two hawks scolding each other from trees several yards from each other. A red-tailed hawk and a goshawk vied for a nest site. The previous year I'd located the goshawk on the nest; later I would return to that tree and learn that the red-tail had won the fight for the nest that season. As I reached into my daypack for my lunch, a wave of wonderment overcame me that I was being paid to hike up here and experience this amazing scene.

I was in college from 1989 through 1991, and in 1992, got my first district wildlife biologist job on the Eldorado National Forest in Pioneer, California. This was the next big personal trial because it meant asking John to leave a place he had never intended to leave, and this was where the shift to my being the main wage earner happened. And it meant leaving Brian behind — he was in college in Sonora.

From there I promoted to wildlife biologist, ecosystem manager and ultimately, district ranger. I didn't like the long-distance parenting, but I did like having the resources to help Brian through college. Each time I got a promotion, he got one too. John and I are proud now that he is successful, with a family of his own and he is a great husband and father. Once my career change was underway, John supported me every step of the way, taking whatever work the then-current community had to offer, which on a few occasions was seasonal work for the Forest Service. He had some hopes for a while of getting on permanently with the Forest Service, but we never became that dual-career couple. For me, having that support was what made what I did possible. Having a husband who changed and grew with me, then supported me, helping with the difficult decisions about the next move, and fully sharing the experiences, has been paramount.

District wildlife biologist responsibilities were similar to what I had done as a biological technician, with the added duties of program planning, budgeting and administration of the district wildlife program. I planned work and hired and supervised employees. I coordinated surveys for Threatened and Endangered (TES) species. I planned and implemented habitat improvement projects. I worked with timber management personnel in planning timber sales. I was on ID Teams for timber sales, vegetation management and other activities. I wrote environmental assessments analyzing activities of other resources for wildlife-related projects. I wrote biological evaluations and consulted with the U.S. Fish and Wildlife Service.

After a year in that position I moved for a promotion to the Daniel Boone National Forest in Somerset, Kentucky. My administrative and IDT duties were the same. The culture and the ecosystem were entirely different. I managed recovery efforts for the federally Endangered red-cockaded woodpecker, and surveyed caves for bats and cliff lines for rare plants. I found I had to slow my "get 'er done" work style down quite a bit, as the people I encountered were in no hurry. It was common to stop and engage in long conversations about life in general with landowners whose property I needed to cross to get to survey sites. Many of those folks had never been as far as fifteen miles from home, but were eager to share their rich histories. One hauled out his *old* Sears Roebuck catalog and another elderly man shared an old newspaper clipping featuring a copperhead snakebite he had survived when he was four years old.

Community interaction involved developing partnerships with volunteers and organizations. I worked with the forest botanist and TES specialist and the local Quail Unlimited to develop a pine-grassland ecosystem restoration project. I represented the Forest Service at meetings of local organizations and developed and presented interpretive programs to various groups.

I worked hard to prove I could do the job and was rewarded when one of my male employees remarked to another, "There's no quit in her." The best was when I left the district and one of the guys at my going away party said, "We hated when we heard you were coming, but now we hate to see you go."

My next assignment was zone wildlife biologist for the Mesa and Cave Creek Ranger Districts on the Tonto National Forest in Arizona. Again, my administrative duties were the same, plus I participated in district organizational planning.

The Sonoran desert was another new ecosystem. Rattlesnakes were common and cactus was treacherous, as was the blistering sun. Forays into my 1,000,000-acre jurisdiction were often on horseback or mule (Figure 174), and occasionally by

Figure 174. Lauren Turner in the Superstition Mountains.

Figure 175. Lauren (front) kayaking Verde River, Arizona to assess southwest willow flycatcher habitat.

ATV. To analyze willow flycatcher habitat, a group of biologists including me kayaked a stretch of the Verde River for four days (Figure 175).

Partnerships with volunteers and community organizations included coordinating activities with Native American tribes with lands adjacent to national forest.

I took on collateral duties as Federal Women's Program Manager, coordinating educational programs, conducting workforce assessment, recruitment and career counseling and EEO type counseling for women.

Most fieldwork did not occur during summer in the desert, so I filled part of that time by learning to do support dispatch as part of the fire fighting militia. Dispatch assignments took me to different locations for three to four weeks at a time. I loved being part of the fire fighting community in that capacity, and now during retirement I still serve as a support dispatcher during fire season.

In 1999, I was promoted again to ecosystem manager on the Feather River Ranger District of the Plumas National Forest in Oroville, California. That was my hardest job ever. Being a district ranger after that was easy. I was a primary staff officer for the district ranger. I led the ecosystem management department on the district, comprised of 35 permanent, full-time employees and at least as many seasonal employees. I directed a demanding program, with ambitious vegetation management targets set by Congress. I oversaw vegetation management, botany, wildlife, heritage and watershed activities, and coordinated support of recreation, lands, minerals, fire and engineering programs.

I was a team member of the forest ecosystem management group, one of the "aunt-ems" that Molly spoke about above. We leveled programs and shared personnel. We made recommendations to top management about program and workforce issues. We became the best of friends.

During my eight years as ecosystem manager, I served in a number of details for about four months at a time, including a forest level assignment as a resource program manager. I had the most fun in in a detail leading my district's recreation, lands and minerals programs. Details as district ranger prepared me for a permanent ranger job.

We moved to Bartow, West Virginia for my first district ranger job on the Monongahela National Forest. This time John left behind a stained glass business he had developed for the past five years.

Bartow had a population of about 200 and quite a bit going on. I supervised staff in minerals, wildlife and range, vegetation management and business management, and shared other employees with an adjacent district and the supervisor's office. As a member of the forest leadership team, I helped set policy, direction and determined program of work and budget allocations.

Developing partnerships that could help communities was what I loved most about being a ranger. I discovered a passion for living in a small community while being able to work toward maybe making it a little better.

My district was located along the Staunton-Parkersburg Turnpike, a National Scenic Byway, run by the Staunton-Parkersburg Turnpike Alliance. We worked together to improve and develop recreational facilities within the national forest portion of the Turnpike.

I worked with the Appalachian Forest Heritage Association (AFHA) to develop interpretation of local forest history. I sub-sponsored AFHA-sponsored AmeriCorps volunteer crewmembers that did resource work on my district and other forest areas.

I sponsored an AmeriCorps VISTA volunteer, whose assignment was to increase economic capacity, and reduce poverty in the community. Together we developed a native plant pollinator garden at the ranger district. We envisioned the garden as both a tourist attraction and an outdoor classroom. The VISTA volunteer developed a network of partners, volunteers and educators to work with the district.

My selection in 2008, as the district ranger on the Hebgen Lake Ranger District back in West Yellowstone, Montana held special meaning for me. That was where I had begun my Forest Service career as an information receptionist in 1980. It speaks to the opportunities available with a Forest Service career, to have come full-circle.

I had the same administrative duties as in West Virginia. My highest priorities remained maintaining internal and external relationships and community collaboration. My district was located along the Gallatin Highway, adjacent to Yellowstone National Park; I coordinated with Park Service employees on issues pertinent to both agencies. I participated with Chamber of Commerce activities, and helped them apply for an AmeriCorps VISTA volunteer to help increase the economic capacity of the town.

A diagnosis of multiple sclerosis was the reason that I retired when I did. The progressive disability struck late in my career, fortunately, and I was able to work for thirty years before I retired. I would have worked years longer, but I became less and less able to be in the field. Though it was possible to delegate, there is no substitute for seeing what was going on, first-hand. As that became less possible, I decided to retire.

I am grateful for having had the opportunity to live and work in so many different beautiful environments. I loved having the chance to get to know different ecosystems and different social cultures. I think that growing up in Yosemite engendered an expectation that life should be full of drama and beauty. The places and experiences in my Forest Service career fulfilled that expectation.

Chapter 9

Retrospective and Prospect

Early pioneering women who worked within the Forest Service paved the way for the women featured in this book who are examples of the many who have steadily continued to help women integrate into the still male-dominated Forest Service. As a result of this evolution, women comprised thirty-eight percent of all of the Forest Service's more than 30,000 employees in 2012 and they now hold positions at all levels and in all disciplines within the agency.[50]

The stories in this book characterize women who have chosen natural resource management careers. A theme that resonates is that these women share a love of the outdoors to the extent that that is where they prefer to work. They value physical fitness, which allows them to actively interact with the land. Their affinity for the land leads to a desire to care for our public lands, where work often feels like play. That desire overrides the difficulties, both physically and societally, of being a woman in a man's world. As one of the major land management agencies in the country, the Forest Service offers a broad range of career options. Women and men can choose to develop their knowledge and skills among a spectrum of natural resources or to specialize in one area.

The Forest Service responded to the forces of change by creating a more welcoming work environment for women. Flexible schedules, virtual workplaces, cross training, family leave and assistance to couples with dual careers now support employees, albeit these accommodations have often come in response to legal mandates. While the civil rights and environmental movements helped more women into Forest Service careers, the agency has needed, and still needs to keep pace to maintain and advance equality in the workplace. Positive changes made to meet legal mandates are frequently followed by periods of backsliding.

One woman in this book postulated that Affirmative Action was important to establish a baseline. The numbers of women and minorities brought into the workforce via Affirmative Action met legal standards without addressing cultural commitments beyond, though the presence of women and other minorities in the Forest Service has had a large effect on the culture. She said, "Today from the local ranger stations to the Washington Office different kinds of clothes and different skin colors are part of the norm. Gays and transgender employees in the small towns where the Forest Service is have agency support." In her experience the Forest Service is ahead of other land management agencies that are still more male-dominated and less comfortable with diversity. And for all its pros and cons, she feels that the consent decree was important, "breaking open the container so what has happened could happen." Women in their late twenties and early thirties are grateful for practices that foster their advancement, but also experience some of the same issues as earlier generations of women holding them back. Others with longer careers have similar observations. One said some of the challenges for women in the Forest Service today are much the same as they were when she started thirty years ago. She thinks there will always be men who cannot respect women as their equals, and challenges and sexual harassment between men and women working together still remain. There will always be young people who come in and have to learn how to behave or not behave around the opposite sex. She said, "We work closely in situations that place you physically and emotionally close with members of the opposite sex. We depend on each other and share experiences that many people never get to experience. We let down our guard because we work so closely with each other. There's a trust level that builds that makes us vulnerable to our coworkers. All of this opens the door to crossing the line between the personal and the professional."

While many women in this book faced challenges of gender discrimination and/or sexual harassment at work, none reported instances of harassment that rose to the level of criminal behavior, but other women in the Forest Service have reported such incidents, including assault or rape.

On December 1, 2016, the House Committee on Oversight and Government Reform held a hearing to examine sexual harassment and gender discrimination within the U.S. Department of Agriculture, including the Forest Service. USDA Assistant Secretary for Civil Rights, Joe Leonard, U.S. Forest Service Deputy Chief of Business Operations, Lenise Lago, former Forest Service employee and current Vice President of the USDA Coalition of Minority Employees/Federal Employee Advocate, Lesa Donnelly and Forest Service Fire Technician, Denise Rice, testified at the hearing. Members of the Committee concluded that the number of cases filed over a period of forty years indicate a culture within the Forest Service. They found that the agency has a poor record of investigating allegations and of holding offenders accountable. Ms. Rice testified about her experience of being sexually harassed, and Ms. Donnelly spoke mostly about shortcomings of the Forest Service in addressing the culture. Mr. Leonard and Ms. Lago reported on improvements and programs that have been underway and that are intended to continue. At the end of the two-hour hearing, members of the committee expressed their commitment to addressing the systemic issues that had been discussed.[51]

Even so, in March 2018, the PBS News Hour ran a story about continuing harassment and retaliation for reporting harassment still being experienced by women in the Forest Service, particularly in fire, which is still the most male-dominated sector of the Forest Service.

Tony Tooke, who was appointed Chief of the Forest Service in August 2017, resigned in early March 2018, amidst allegations of sexual harassment. Upon his resignation, he issued a letter to Forest Service employees, in which he neither clearly denied nor confirmed the allegations against him, but stated that he had requested, fully supported and was cooperating in an investigation into his past behavior. Vicki Christiansen was named Interim Chief upon his resignation, and was tasked with improving the agency's response to sexual misconduct.

Tooke began his letter by saying that every Forest Service employee deserves a safe and respectful workplace free of harassment. He acknowledged that recent women's stories of sexual harassment in the Forest Service reveal that the agency still has much more to do to achieve a safe, positive and respectful work environment, and that leadership is committed to investing in changes and resources needed to bring about the cultural change needed, to address the drivers in our culture and change the systems that allow harassment, bullying and retribution to occur.

The recent attention to sexual harassment and misconduct throughout the Forest Service and other federal land management agencies prompted James Lewis, author of *The Forest Service and the Greatest Good*, to reexamine what he had written in his chapter originally titled "New Faces, Changing Values," which had examined the evolution of the roles of women in the Forest Service. Lewis said that if he had written the chapter today, he might have titled it "New Faces, Same Old Values," since all in all, not much has changed.

While the Forest Service has made some progress over the past decades, it has been in fits and starts with one step forward, two steps back, pretty much in lock-step with overall societal trends of the day, mimicking what is going on in the culture of our society as a whole. Perhaps the contemporary "Me Too" movement and "The March for Our Lives" offer us hope that we are ready to make significant forward gains again. Women are insisting on day lighting the entrenched culture of sexual harassment that permeates business, sports, entertainment and the government, and they are demanding accountability. Young people, raised by these women of a different mindset and their partners, are poised to take on strong leadership roles and they are demanding change. It feels like a renewal of the kind of activism that brought about the momentous societal changes of the 60s, 70s and 80s. There is reason to hope for a future that will achieve what Tony Tooke said is needed to effect real cultural change in the Forest Service.

Both young and more seasoned women have shared their points of view with me about pros and cons during their Forest Service careers, and what they think the future may or should bring. I summarize those perspectives below. Their perceptions are as varied as the women, but in many cases their experiences are similar. I have consolidated like comments. To facilitate open sharing of these ideas, individual respondents are not identified.

There was a sense that despite significant changes, women still have to perform at higher levels to get the same promotions and recognition, and do another shift at home. Some feel that the pressure to prove oneself was greater when women first started entering the Forest Service, and that the greater challenge has shifted to career-personal life balance. For women, generally more than for men, having a family can still conflict with growing a career. These changes have presented challenges for the agency in figuring out how to keep qualified men and women while allowing flexibility in their personal lives. Families still have conflicts, but due to accommodations by the agency and the evolution of society, there is better acceptance for variations in how careers are managed and how family work gets done.

In some male dominated fields within the Forest Service, women still practice masculine behaviors in order to fit. These are fire fighting, trail and timber crews, heavy equipment operators and other fieldwork. This is most prominent in the technician and law enforcement arenas.

One woman remembered lots of women in fire leaving the Forest Service in the 1970s and 1980s. One line officer said that she had not heard about the agency as a whole having issues with retaining newly hired or young women [circa 2014]. She was aware that in Region 5, the fire organization was struggling to retain women. She said, "My very unscientific observation of this is that many young women choose to have kids and that, at least while the kids are young, puts a crimp in those open-ended fire jobs that involve lots of time away from home, and odd hours. There are also isolated issues with sexual harassment that do not always give the organization/fire a good name."

Women smokejumpers remain a minority in the fire sector. As of 2003, of the nation's more than 400 smokejumpers, twenty-seven were female and the proportion has actually decreased since.[52] According to a 2018 poll provided to me by Jennifer Jones, Public Affairs Specialist at the National Interagency Fire Center, only twenty of approximately 445 smokejumpers nationwide were women. One woman who contributed data to that poll said, "I just have to add: this [the small number of women] is not because women aren't accepted, it's just because there are a lot less women fire fighters who apply to be smokejumpers. When women apply to the smokejumper programs they are hired in exactly the same manner as male candidates." Despite the relatively small proportion of smokejumpers that are women, smokejumpers seem to feel that sexism does not play a big part. Women smokejumpers feel camaraderie with their male counterparts, and refer to them fondly as "the bros." The men likewise refer to their female counterparts as the bros. One female smokejumper thinks the biggest issue for women getting into jumping is the physical requirements, mainly upper body strength. Chuck Sheley, Managing Editor of Smokejumper Magazine, and a former Forest Service trainer of thousands of wildland fire fighters maintains that the upper body strength requirements are actually discriminatory towards women. He examined the issue at length in a 2015 article in the magazine, "Is The Smokejumper Physical Fitness Test Eliminating Good Women Firefighters From Smokejumping?[53]

The same woman who said she thought the upper body strength requirement was the biggest issue for women getting into smokejumping said, "There was a time when seeing women in smoke jumping was not the norm. I don't think that the younger jumpers even think about it. It is now normal to have women in the organization."

There is evidence that, although still a minority, women smokejumpers are now considered a normal part of that field. There is an annual gathering of smokejumpers from all over the country. For the 2013 event, organizers included in the brochure that announced the reunion an invitation to men and women smokejumpers alike and to those "who are or ever were spouses and partners, pilots and all those who ever stuck a Pulaski into a burning log or knew or worked with those who did," to celebrate thirty years of women in smokejumping. Stories were told about "firsts," the first women to break ground in smokejumping. A stellar moment was when they realized that while there were some fairly recent firsts, it is not such a big deal; women now "are it," a normal part of the smokejumper workforce. This was ironically borne out by a woman honored as the first woman to retire after twenty-seven years as a career smokejumper.[54]

Women who choose smokejumping as a career face the same challenges around having and raising children as other women, perhaps a bit more profoundly. There have been pregnant smokejumpers, women who juggled smokejumping with motherhood and those who have chosen to be stay-at-home moms, at least for a while. There is still a worry that if you take a break, it will be hard if not impossible to get back in. That issue is not exclusive to smokejumpers. A woman from another discipline said that she knew women who left the agency due to the lack of flexibility of their male bosses towards their having children. Those women waited until they had well-established careers to have children. Another woman told the story of a district ranger she worked with early in her career. She was trying to transfer to his district for personal reasons. The ranger called to interview her, and he asked her if she had kids. When she said no, he told her that was good, because there were issues with women taking time off work to take care of sick kids. She ended up getting a different job in the supervisor's office on that same forest and found out that there were a bevy of stories about that ranger and that kind of thing he did, making her very happy to not have gotten a job on his unit.

She discussed at length the difficulty of raising children while working for the Forest Service:

> Some of it was about women, but most of it is applicable to family situations. Most folks have kids at some point, and that injects a lot of challenge (and joy!) into life. Kids get sick, usually Mom stays home to take care of them and working that into a tight schedule at work is not fun. Childcare is also a big issue, especially in small towns. You can't find reliable, quality childcare. It's tough to find childcare that can accommodate long work days or travel. There are more single female than male single parents. Often, due to childcare, women aren't able to attend the same trainings as their male colleagues. Working through the dynamics of a two wage earner family is tough. The feeling of having an overflowing plate at home and at work can be pretty overwhelming. Most of what I am talking about is not unique to the Forest Service. The Forest Service is just a reflection of what is going on in greater American society.

The fact that up until 1978, when the Pregnancy Discrimination Act was passed,[55] a woman in the United States could be fired from her job for being pregnant[56] tells how far we have come. The United States is the only industrialized country that does not have mandatory paid maternity leave.[57] Forest Service employees can use sick leave. Women smokejumpers and non-smokejumpers have worked out support systems, such as a supportive spouse at home, or other family members, such as a grandmother, caring for children while she was on assignment. Some dual-smokejumper families have alternated their fire assignments so that one of them was home with the kids. Many of the smokejumpers that attended the 2013 reunion brought their children with them.[58]

The Forest Service has for decades sought ways to be an employer of choice, inclusive of women and minorities. Agency demographics fluctuate, in response to budget constraints and other factors. Agency records show that in 1988, women comprised 33 percent of the Forest Service workforce, and minorities represented 13 percent. In 1992, there were 40 percent women and 15 percent minorities. Representation fell to

38½ percent women but rose for minorities to 17 percent in 2011.[59] Data from 2017, show women still at about 38 percent and an increase in minorities to 20 percent.[60] One woman who entered the agency in late 1980, and who tracks agency demographics has borne witness to these ups and downs. This woman thinks that the agency can do more in diversity hiring. She said each chief has priorities, and cited Chief Dale Robertson's Workforce 1995.[61] In her opinion he was the best since she had been in the agency to make diversity a commitment. She was referring to a concerted effort under Chief Robertson's leadership to create and maintain a diverse workforce. That program made significant progress in its first decade, which later slowed.

Another woman cited circa 2013 data that showed the gender mix throughout the agency being about what it had been for more than twenty years, also showing stagnant numbers in ethnic diversity, which is reflected in the data above. An increase in diversity from 13 to 20 percent over a period of nearly thirty years (1988–2017) is fairly stagnant. She pointed out that the Forest Service was making a considerable effort getting their recruitment efforts to reach a broader spectrum of people.

A woman working on a detail in 2012, that examined representation of women in the workforce, heard Forest Service Chief, Tom Tidwell, say that women were leaving, and that fewer were coming into the agency. Agency data show a reduction of both female and male employees between 2011 and 2017, with a reduction of the proportion of women of one percent from 2012 to 2013, and reduction of an additional one percent of women from 2014 to 2015. The percentage of women then remained steady at 34 percent representation through 2017.[62] Chief Tidwell issued a letter to employees during Women's History Month in March 2016, citing a new effort to revitalize and modernize the Forest Service Women's Special Emphasis Program to "create a work environment that better reflects the needs of women."

Another important demographic changed in the Forest Service since the 1970s, when women started entering its workforce in numbers at the same time that the agency started to expand the sphere of specialties they were hiring to include disciplines other than forestry. One woman cited a downward trend in forester positions between the late 1990s and 2012. Agency data show a reduction in the number of forester positions in the Forest Service nationwide from 3,297 in 1998, to 1,603 in 2012, and a further reduction to 1,319 by 2017.[63] While she found the downward trend disturbing, and it is an issue for the forestry schools that train foresters, the majority of women who addressed the change viewed it as a positive transition. Another woman said that there are fewer females and people of color studying what one woman called "traditional natural resources." She said more are studying environmental science, conservation biology and environmental biology while the agency is still tied to the traditional natural resource careers. She linked that to the current-day trend of a large number of people residing in urban areas. She thinks the focus needs to be more on the connection of people to nature, not so much on strict forestry. She tells people that she is a forester and we still need foresters, but she would rather hire someone with skills in communication, information technology, collaboration, social science, problem solving and inter-relationships than in traditional forestry. A woman from another specialty said that one of the challenges she sees affecting both men and women is the Forest Service's lack of training the newer generations in the specialty positions. She said it is getting very difficult to fill technical positions at higher levels.

Another said, "Most women are in non-timber positions which I think must have shifted the identity of the Forest Service more towards recreation and ecology." To illustrate this point, agency data for 2017 show 336 women in forestry positions. Data for women in a sampling of other positions in that same year show 640 women in ecology, botany, soil science, fish biology and wildlife biology, and 1,007 women in general natural resource positions, which recreation management positions often fall under. Of the 3,297 forester positions in 1998, women held 24 percent. That proportion increased to 26 percent in 2012, and remained the same in 2017.[64]

One woman looked to the societal and parallel agency evolution that began during the early seventies. She said, "The effort to incorporate women and minorities into the workforce and allow them to advance in their careers has meant that the Forest Service has had to look beyond the traditional. As an example, it used to be that if you wanted to advance up the ladder from a field staff to a district ranger, to a forest supervisor, to a regional forester and on up, you had to be a forester. Any others — wildlife folks, entomologists, human resource specialists — should forget about it. To get women in positions such as those the agency had to look beyond traditional forestry. I remember when a computer scientist (a man) was selected for a district ranger position — what a concept that was! Now we have even had a woman with a human resource background as a regional forester."

Another woman noted the need for varied skills within the Forest Service, including good communication and information technology skills, noting that the agency continues to find challenges in maintaining contact with the public. She said, "There is a disconnect between many Forest Service employees and the resource work they do, and being able to connect with, interact with and find ways to share the agency management issues and mission with the public." Another said, "If we don't communicate what is significant, someone else will, with their own spin on it." She viewed the Forest Service to be in a precarious position with reduced government spending resulting in less *facilities*, *services* and *staff*, three elements that have historically been used by the government to connect with the public.

One woman thought changes over the past several years have impacted the overall "personality" of the agency as an employer. She was first hired directly from her college through the STEP program. STEP was beneficial to college students who needed three months of summer employment, and it was easier to try out a job for a summer. She thinks the program encouraged a greater variety of personalities, experience, education and gender in employees.

Once hired as a temporary, an employee with satisfactory performance can be rehired each season without further competition, so this young woman's serial temporary appointment was secured after her first season. Her challenge then became securing a permanent job. She thinks that most people who stick with it are rewarded with permanent status, though it may take several years. She quipped, "It took my supervisor almost twelve seasons! I figure I have five more years before I can really complain." As mentioned in Chapter Two, the Land Management Workforce Flexibility Act has alleviated the situation for long-term seasonal employees. More details about that act were described in Chapter Two and are reiterated in the discussion below, related to Veteran's Preference.

A breakdown of the career-ladder system in fire has made it hard for new recruits to advance once they have gotten a job. Fire technician jobs generally range from GS-3 through GS-9, and require a year of experience to promote from grades GS-4 and above. Region 6 promoted engine assistants to GS-6 and engine captains to GS-7, which left a gap at the GS-5 level due to the fact that most seasonal employees in fire only go to the GS-4 level, and cannot jump from GS-4 to GS-6 without first working at the GS-5 level.

Another, non-fire woman noted a lack of career ladders overall, which is making it harder for women and men alike to advance in Forest Service careers. She rued the loss of the traditional career ladder, saying, "Nothing beats having the knowledge and skills that comes from working your way up through the organization."

Along with this break in the ladder, Affirmative Action has had a great impact on hiring practices in the last few years, which one woman believes has pushed minority hiring to an extreme. She said, "It is virtually impossible, even in places where there is almost zero diversity, to be hired as a white male, and almost as hard as a white female. The Forest Service used to consider women to be candidates for 'diversity' hiring, but now it seems that hiring women is not receiving the emphasis it once did." Note that since the time of her comments, the Forest Service has again stepped up its efforts to recruit women, with some emphasis placed on hiring women into fire fighting jobs.

Another hurdle for women is that most new hires are military veterans; jobs must be offered to veterans on hiring certificates before anyone else on the list can even be considered. This woman sees this as having one of the greatest impacts on the number of women in the Forest Service, now and in the future, as most veterans are not women. She has personally been passed over for permanent positions that she was well qualified for, in favor of veterans, and she knows other women who have had that experience.

By law, veterans who are disabled or who served on active duty in the Armed Forces during certain time periods or in military campaigns are entitled to preference over non-veterans in hiring. The goal is not to place a veteran in every vacant federal job, but to give a leg up to those have served our country at significant personal sacrifice.[65]

Clearly veteran's preference will always be an issue for women or others who are not eligible for preference. This is speculation on my part, but it stands to reason that if the economy and budgets improve, competition for all jobs would ease. True equality in our society may come as more women serve in the military. The Forest Service has developed The Women and Veterans in Fire Training Program, a unique employment and career preparation opportunity for eligible veterans interested in wildland fire fighting.[66]

The Land Management Workforce Flexibility Act, enacted on August 7, 2015, allows people who have worked as temporaries for the Forest Service and other federal land management agencies for a total of more than 24 months without a break in service of longer than two years to apply for jobs under internal merit promotion procedures, previously only open to permanent employees. This opens up a lot of job opportunities, and also puts all applicants on an even playing field, since veteran's preference does not apply to merit promotion. It does apply to the temporary jobs these employees originally compete for, and as such it shifts competition with veterans to initial temporary appointments.

Some women are able to get into the permanent workforce by first applying for clerical jobs, which are traditionally occupied by women, and then transferring to their desired position once they have secured a permanent position. One woman said that while she sees the value of the tactic of jumping into a clerical position in order to get permanent status, she also sees that there are still a disproportionate number of women stuck in clerical positions in the Forest Service and she hopes to buck that trend.

Instead of focusing only on diversity, she thinks the focus also needs to be on quality of employees and not just filling quotas. Another woman's opinion that working with the Forest Service is not for everybody aligns with that thinking. She said, "I've worked with some amazing women and I've worked with some women that, looking back, I wish I'd never had the pleasure. My experience has been mixed. Some of the greatest supervisor's and coworkers I've had have been women. Some of the worst have been women. At times I think that the Forest Service may have sacrificed some of its integrity in trying to bring more women on for the sake of having more women in its workforce. I'm not sure the right women were always brought on."

Declining budgets have played a large part in contributing to loss of gender and diversity in the Forest Service. There has been an overall reduction in the number of employees in the agency. In 2010, the agency had over 42,000 employees in the United States, which began a decline in 2011 to 40,436. Trend data for the years 2013 to 2017, show a further decline from 38,094 in 2013 to 36,358 in 2017.[67] A large number of women and people of color came into the agency in the 1960s, 1970s and early 1980s. They are now retiring and not being replaced due to a reduced emphasis on outreach and recruitment, which is related to reduced budgets. One woman noted that the agency is left with a huge number of employees in their forties and fifties who are not ready to retire, while younger employees are not interested in a GS-5 with no promotion potential. She feels the agency has to figure out new ways of bringing the agency elders and the new hires into work environments and mentoring scenarios. She said, "The agency has created a new hiring system to recruit young college trained professionals [Pathways][68], but our systems work slowly, and the current environment for hiring equals nearly seventy-five applicants for one position." She believes that creating working teams made up of different ages, grades and disciplines is actually as important for agency transformation as the work itself. She said, "We learn from each other, and my experience with college students always leaves me feeling hopeful and optimistic about things to come."

Young people are also scared in this economic and political climate to take a job that is dependent on public funds, doesn't allow for year round employment and is not a "power" job. One woman said, "Most of my friends either went on to grad school, or immediately into accounting/econ/business, all considered 'safe' careers when we graduated in 2007. They still have a very hard time understanding, first, what I do and second, why the hell I would do it (risk-wise when I was on a fire crew and employment-wise in terms of being laid off seasonally)."

Reduced budgets have impacted the ability of the Forest Service to offer attractive benefits to employees. Transfer of station packages that help employees relocate for promotions have diminished or are not offered. Spousal placements that provided a safety net to dual-career couples have diminished. It is, however, still much improved over the early days in the 1970s and 1980s. One woman told of one of her challenges of being a dual career couple in the Forest Service during that time. She said, "The system was set up to help the men achieve their career goals through promotions and moving to different locations. The wife was expected to follow. My first marriage didn't last because of this system."

The Forest Service later made efforts to assist dual career couples, but today many positions that were great counterpart jobs for couples with one person in fire and the other looking for a job that required a somewhat lesser time commitment have been cut.

Those seeking high-level senior jobs in the agency face the same issues. One woman said that typically in order to advance beyond a GS-13, an individual must singularly devote themselves and their time to Forest Service work. She thought that few families would choose to move to Washington D.C., where a lot of the higher graded positions reside. Her perception was those employees who go there are single, single career couples, or are empty nesters. Many forests do not hire from within their own staff, so families are left to choose disrupting their entire family for one person's career.

Another woman remarked that she thinks the challenges for women in the Forest Service are the same as they are everywhere: they tend to be the ones expected to care for the children and household and also hold down a job. Also, until the expectation for people to move in order to move up is abandoned, the "second earner" in the family (often the woman) will usually have to sacrifice her career for her partner. This woman said she was fortunate in that she was the one with a career, and her husband followed her around the country. She said, "I'm seeing some change, but we still have a ways to go before career ladders in place will be the norm." A related opinion was that the Forest Service should consider family dynamics in job placement and to continue to encourage couples to pursue careers within the agency.

A mother of five said it was exceedingly rare for women in leadership roles in the Forest Service to have as many children as she does. She said, "I actually do not know any women leaders, out of all the ones I know, who have five children. The majority have no children, and the others maybe two, never more. That was how it was to get that far up in the male dominated ranks of the Forest Service. God help you if you were trying to raise children at the same time you wanted to get ahead." She said many of her women friends forfeited the privilege of being a mother to ensure their roles were secure or to climb higher in the male dominated ranks of the Forest Service.

Likely tightly related to reduced budgets and increased competition, some women voiced concern that the Forest Service culture has become more "cutthroat" because everyone who is employed or who even hopes to start a career with the Forest Service has started to see their coworkers as competition instead of allies. One said, "When there are so few jobs and such little support for employee growth, you have to look out for Number One. This attitude deters some great people from pursuing careers."

A similar sentiment came from another woman who said that in her opinion the law enforcement and investigation arena in the Forest Service had taken a long step backwards in the past few years. In sharp contrast to when she first started, it had become a "good ole boy" place to work. About twenty years into her career her unit got an inexperienced new captain, who after two weeks on the job tried to terminate her, with the help of his female supervisor, based on a complaint from a private citizen that she said they knew to be a false complaint. It was a lengthy process, about which the woman said, "Thank God for our union." The captain did not succeed in getting rid of her, but treated her like a second-class citizen during her last three years. She said, "Whether you win or lose, it takes a toll on family life and health when you have to fight. While I mostly loved my Forest Service career, some of the smiles hid a lot of distress. I loved the Forest Service until the last four years of my career. Bad management and people in positions who had no business supervising changed me from a loyal proud employee to a person who just *made it to retirement*."

Another woman under that same captain, also with about twenty years under her belt, took a voluntary early retirement to escape the situation, and that resulted in her losing her law enforcement retirement supplement. There were a total of four women in that region alone who had felt pushed out in the end

— half of the female law-enforcement workforce in the region within four years.

One of the women who spoke of a cutthroat environment throughout the Forest Service said that she felt those left who have ethics and a moral high calling to standards and value-based thinking get shunned as dinosaurs and are considered obsolete. Individual employees are treated as though they are dispensable commodities — units — and it is okay to disparage, shun and excommunicate those who "think differently," think outside the box or shine in a way that is "not the quiet girly voice for women leaders, the head down non-confrontational demeanor they have homogenized into employees in general." She thinks this was also evident when the agency chose to call human resources "human capital" which was rapidly changed back to human resources because the term human capital was so demeaning to employees. She is referring to an agency move, again in response to reduced budgets, when administrative functions across the Forest Service were merged and many positions moved to Albuquerque. The agency took two hits as a result of that move, loss of positions and loss of females or people of color who had been in those positions.

This woman also believes that tight budgets and intense competition have hurt morale, negatively affecting productivity and threatening the Forest Service status as an employer of choice. Under the George Bush mandates of outsourcing and downsizing from the 33,000 permanent employee base that had been stable throughout this woman's thirty-year career, the Forest Service set out to cut more than 10,000 permanent employees, resulting in infrastructure that is falling apart, forests that have been regenerated and are now un-managed and thousands of overworked and underpaid employees. As the workforce diminished, the work did not. The Forest Service is responsible for 193,000,000 acres of forested lands that provide twenty-five percent of the clean water for this country, with millions in backlog maintenance on miles of recreation and logging roads, historic buildings crumbling and treasured trails in disrepair across the country.

Still this woman says that she is eternally grateful for her work. She said, "I hurt for the agency I love — this is a job I have loved every day of my entire career of over thirty years now! I still cannot believe I get to work in the forests I love so much."

Several women acknowledged that while the entrance of women into the Forest Service had changed the agency, their careers in the Forest Service had also changed them, most often in positive ways.

One said she thought that working for the Forest Service gave her self-confidence, because of the challenges she experienced — both systemic challenges in the early days and career challenges presented to her by her "wonderful supervisors." She said, "I attribute my success to (a) good supervisors who gave me opportunities, (b) my own willingness to say yes to new opportunities, and (c) my stubbornness, persistence and the willingness to try again because I loved the work and the land."

Another said that she certainly has learned more about resource management and sustainability working for the Forest Service than she would have elsewhere. She said, "I've had more opportunities to explore things I enjoy, such as accessibility and leading education efforts, than I probably would have in civilian life or most other agencies."

Another woman said, "As my career progressed, it became clear that I was doing a job beyond what I signed up for. I liked the work, but also serving the public in that capacity. I grew up in the Forest Service and many great role models along the way provided learning opportunities and challenge to make me a better professional."

Other women overwhelmingly felt that despite its faults or challenges for women, overall the Forest Service is a great agency to work for. One said she was contemplating leaving the organization about fifteen years into her career. She decided to get her masters degree, thinking of maybe working at the university. She came to realize that the university had the same negativities that the Forest Service did, maybe even worse, in terms of competition, resentment, negativity and the like. So she said, "I decided to stay and to be as positive as I could and hopefully make it a better place around me."

Another felt that the federal agencies for the most part, have come light years in recognizing women and their contributions in the work place. These women love their jobs, and the opportunities to travel and expand their knowledge and skills.

Another placed Forest Service ahead of other agencies for fostering constructive criticism and improvement and cohesion. She said that both female and male employees are proud to work for the Forest Service and they still share the sense of camaraderie among all Forest Service employees. Another said that no matter how crazy the environment was, in the end every office she has worked in has been like a family. She said, "We may fight like cats and dogs internally but most people would defend you to the core when it came to the outside world." She went so far as to say that since she has worked for the Forest Service for her entire career she feels that the agency "grew her" and she is who she is today because of it.

Another repeated that what got her into the Forest Service was her passion for the forest and deep belief she could make a difference, her passion for our public lands and wanting to work in the woods. For all the angst she sometimes feels, she finished by saying that she has loved every day of her career with the exception of the past few years in which budget-driven changes within the agency had devastating affects to her. She talked about the change in culture from that of the 1960s' and 1970s' "antiwar influence, save the earth, world peace issues." Her generation was part of the women's emancipation movement that saw women being liberated after centuries of subservience in the United States. Her generation fought the battles and reared a generation that does not have to give a thought to what were big issues during their parents' day. They can take it for granted to a greater degree than their mothers that they will be equals in the workplace.

Another woman said she was pleased to see the progress women have contributed to in gaining acceptance of people in professional positions regardless of gender, race or creed. As a result the Forest Service has focused on hiring more diverse candidates, and different types of skills. She will not ever tell her daughter she can only be a teacher, a secretary, nurse, bookkeeper or typist, comments she heard from her high school counselor. Another woman stated, "Women were the first wave of minorities to enter the Forest Service and other minority groups such as Hispanics, African Americans, Native Americans and Asians soon followed. The Forest Service today is more culturally diverse than it was twenty-five years ago when I entered the agency, and much of that can be attributed to the strong women who paved the way for it thirty years ago."

Reflecting back on her entire career, another said, "Giving women the opportunity to work for this agency has resulted in a broader view of what people can accomplish if given the chance. Women have been part of the Forest Service for many years, but prior to the seventies and eighties, were largely in administrative or secretarial roles. There were a few women foresters, but they were rare. My generation saw the expansion of women in traditionally male-dominated jobs including fire, range and forestry. These women had to be tough and thick-skinned. As women become more common in these positions, I think their true natures are being allowed to show. We are less about having to prove ourselves and more about getting the job done. We are a more diverse organization, with a greater acceptance and tolerance of people's differences and appreciation of what each individual has to offer."

Yet another said, "Women have an equal face in defining the Forest Service now. The same evolution has been happening with other minorities over the last ten years or so and their stories may be very similar to women's stories. I can't speak for all women but I have never felt that I was not welcomed or valued as a woman in the Forest Service. Women have 'stepped up' to the proverbial plate and performed well at whatever job duties were given. Women have helped the Forest Service fulfill their mission — both the vision of an integrated workforce but also their original mission for the land."

One said she thinks that the movement of women into the work force has shifted things for society as a whole. She thinks that women with abilities in science, technology, engineering and math, women who are educated and have an income, drive many family decisions. She speculated that the Forest Service's current emphasis on cultural diversity and inclusiveness could be the next generation beyond just bringing women into the work force. On the other hand, she has known younger women who do not seem to have a similar relationship as she does with shared decision-making. She said, "There seems to be this nuance of here's what guys do, here's the gals duties. Maybe it's a bit of a pendulum swing from all the work of the liberation movement." Another observed that her daughter's generation sees people as equals and gender is less relevant. The whole workforce is valued. She thinks that might save morale in the Forest Service. She ended with, "I will optimistically believe it will because the mission is still one of the greatest on earth, looking after public lands that are so vital to the wellbeing of our country."

One woman pointed out that women had had a dual struggle, both in needing to demonstrate the ability to physically perform the work and being respected in leadership roles. She said that the male culture of the workplace had been dismissive of women in leadership and hyper critical of their performance. Change in that dynamic was observed by another woman who said, "The Forest Service of today is much more accommodating to women than it was twenty-five years ago when I started. Women are not as much of a minority and do not have to search as hard for support. Most men seem to have accepted women in the workplace and many have had female supervisors, which was not the case twenty-five years ago. Women have shown they can be taken seriously and men tend to not be as judgmental now as when they first start working with women."

Likewise, the woman who told the story of the ranger who discouraged women from having children said that guy would never get a ranger job today. She said:

> Our thinking around what we value in leaders today — inclusion, the ability to communicate, working effectively with others, promoting respect, demonstrating self-awareness — is so different than what the same values were thirty-five years ago (overriding objective was getting the work done and holding people to 'the rules'). That is not to say that all rangers back then were like the guy I described, but there were a lot of them. What was valued and rewarded is very different than today. I think that is a big difference and change, and it is one that better accommodates a broad array of employees — men, women, people with disabilities, people who are non-white, people who are non-heterosexual, etc. It matches up with the modern pool of potential employee candidates. Back in the day, searching for just the right white guy who went to the same forestry school as you and aligned 100% with the values you held may have worked just fine, but the world is so way past that. Our agency is so much better because of the diversity that we've brought in. As a result, the ideas and solutions brought forward to meet the challenges we are facing are more creative, robust and responsive. The agency is different, and as a result it is better.

The Forest Service, like any agency or workplace, has its issues. In order to stay relevant in a changing world, it needs to build on past success and continue to connect with the public it serves and the future employees who will provide those services. All of us need the resources that our public lands provide, whether or not we recognize the connection between those resources and the management of the lands that produce them. The Forest Service offers employment to those perhaps uncommon persons who want the outdoor, rural life that a career in natural resource conservation provides, those who cannot accept life without grass, trees, gardens, and wildlife around them.

In 2004 and 2007, Sharik and Frisk conducted surveys of undergraduate student leaders in forestry from around the country.[69] They examined demographic characteristics of respondents and positive and negative factors influencing their decision to enroll in forestry programs. While the focus of this book is not on the status of forestry school enrollments, the findings in Sharik and Frisk's study support observations of the women who contributed to this book. They noted the trend away from pure forestry toward other natural resource studies. They also cited an increasing disconnect, related to increased urbanization, between society, particularly young people, and natural resources related to increased urbanization.

In 2008, Sharik and Lilieholm synthesized the findings from the earlier survey by Sharik and Frisk.[70] They cited limited public awareness regarding social benefits of forestry and natural resource professions. They found that reduction of enrollment in natural resource fields may be attributed to a weak and uncertain job market in the industry, and to lower salaries than other professions. They recommended that employers of forestry graduates raise salaries to levels competitive with other professions. I would point out, though, that for those who do desire work in the forests, I have found the salaries for Forest Service workers to be better than those of other work available in the locations where Forest Service jobs are located.

Sharik and Lilieholm found the primary reasons cited by students for enrolling in a forestry program included a love of nature and a desire to work outside. The love of nature was the greatest attraction. Whether raised in urban or rural settings, childhood outdoor activity and early exposure to forestry were positive influences in their career choice.

The extraordinary women profiled in this book and many like them paved the way for women to enter and find fulfillment in Forest Service jobs once reserved for men. They exposed their children to positive outdoor experiences. By their examples they raised daughters who expect to be free to choose a career based not on their gender, but on their interests and abilities, and sons who take in stride that their sisters may share their same career interests. They have incidentally and incrementally changed society to benefit both men and women. Men now also have more choices to participate in their children's lives and to share the burden of raising and providing for their families.

One woman said, "I don't know how to measure how far we've come, but women are definitely making a change in the way the agency looks at inclusion, diversity and work-life balance. I believe we'll see a great deal more change as the new generations of women continue to show up expecting equal treatment and consideration. If we can wholeheartedly embrace that change, we'll be able to accomplish amazing things."

Another said she has worked with women who could physically outwork many men, and with women who worked smarter than many men. She said, "I think working smarter is probably where women's influence has been seen the most. We use technology, and build relationships and partnerships to help get the job done. Our mission is the same, "caring for the land, serving people." We just do it a bit differently now"

Just like the society it serves, the face of the Forest Service may change, but its core mission, "Caring for the Land and Serving People" is as relevant as ever, as are those who have chosen and will choose as their personal mission conserving our natural resources for the "greatest good, for the greatest number, for the longest amount of time."

For information about applying for jobs with the Forest Service, visit the Forest Service website at http://www.fs.fed.us and click on the Working With Us tab. Another Forest Service publication, The U.S. Forest Service — An Overview, is a detailed summary of facts about all aspects of the Forest Service. It can be accessed at http://www.fs.fed.us/documents/USFS_An_Overview_0106MJS.pdf.

Endnotes

1. Williams, Gerald W. "Women in the Forest Service: Early Historical Accounts." Unpublished manuscript, U.S. History Collection, Forest History Society, Durham, NC, 7 and 9.
2. www.fs.fed.us/aboutus/mission.shtml
3. http://www.blm.gov/wo/st/en/info/About_BLM.html
4. http://www.bia.gov/WhoWeAre/BIA/
5. http://www.fws.gov/info/pocketguide/fundamentals.html
6. http://www.nps.gov/aboutus/mission.htm
7. http://www.nifc.gov
8. http://www.doi.gov/americasgreatoutdoors/index.cfm
9. http://www.fs.fed.us/about-agency/organization
10. Lewis, James G. *The Forest Service and the Greatest Good: A Centennial History.* Durham, NC: Forest History Society, 2005, 43.
11. Steen, Harold K. *The U.S. Forest Service: A History.* Durham, NC: Forest History Society and Seattle, WA: University of Washington Press, 1976, 44.
12. Duhse, Robert J. "The Saga of the Forest Rangers." *Elks Magazine*, July-August, 1986.
13. Williams, Gerald W. "Women in the Forest Service," 5-6.
14. See *What did We Get Ourselves Into: Stories by Forest Service Wives* (Missoula: Northern Rocky Mountain Retirees Association, 2000.)
15. Pendergrass, Lee F. "Dispelling Myths: Women's Contributions to the Forest Service in California," *Forest and Conservation History* 34 (1) (January 1990): 1.
16. Williams, Gerald W. "Women in the Forest Service," 2.
17. Williams, Gerald W. "Women in the Forest Service," 2 and 4-5.

18. Agriculture Information Bulletin No. 402
19. Williams, Gerald W. "Women in the Forest Service," 1.
20. Lewis, James G. *The Forest Service and the Greatest Good: A Centennial History*, 171–172.
21. West, Terry L. *Centennial Mini-Histories of the Forest Service*. Washington, D.C.: United States Department of Agriculture Forest Service, FS–518, 1992.
22. Steen, Harold K. *The U.S. Forest Service: A History*. Durham, NC: Forest History Society and Seattle, WA: University of Washington Press, 1976.
23. Traverse, Hannah. "SheSheShe" Camps: A Women's Alternative to the Civilian Conservation Corps. The Corps Network March 2013. http://corpsnetwork.org/shesheshe-camps-womens-alternative-civilian-conservation-corps. Further sources cited at the end of the article.
24. http://www.thesca.org/about
25. Biographical and Historical Notes, Gene Bernardi Papers, Library and Archives, Forest History Society, Durham, NC, USA.
26. Dawson, Larry J., and Alicia D. Bennett. "The U.S. Forest Service Job Corps 28 Civilian Conservation Centers." USDA Forest Service Proceedings RMRS-P-64. 2011
27. https://www.corpsnetwork.org/about/history
28. www.wilderness.net/NWPS/FS
29. http://www.fws.gov/laws/lawsdigest/esact.html
30. https://www.nationalservice.gov
31. Williams, Gerald W. "Women in the Forest Service."
32. http://www.fs.fed.us/about-agency/organization
33. http://www.fs.fed.us/research/outdoor-recreation/
34. http://www.fs.fed.us/eng/transp/
35. http://www.fs.fed.us/eng/structures/
36. http://www.fs.fed.us/eng/dams/
37. http://www.fs.fed.us/gstc/

Endnotes

38. http://www.fs.fed.us/eng/techdev/
39. http://www.fs.fed.us/eng/fleet/
40. http://www.fs.fed.us/eng/facilities/
41. Ragenovich, Iral. personal comments
42. http://www.fs.fed.us/research/about/
43. http://www.fs.fed.us/about-agency/international-programs/program-topics
44. http://www.fs.fed.us/fire/people/smokejumpers/bases.html
45. http://www.fs.fed.us/lei/organization.php
46. National Smokejumper Training Guide. USFS 2008. Physical Conditioning Lesson Plan. http://www.fs.fed.us/fire/aviation/av_library/sj_guide/05_physical_conditioning.pdf.
47. https://www.opm.gov/policy-data-oversight/senior-executive-service/scientific-senior-level-positions/
48. Clance, Pauline Rose, and Suzanne Imes. "The Imposter Phenomenon in High Achieving Women: Dynamics and Therapeutic Intervention." Psychotherapy Theory, Research and Practice, Volume 15, #3 (Fall 1978).
49. http://www.fs.fed.us/about-agency/organization
50. http://blogs.usda.gov/2012/03/16/u-s-forest-service-women-opportunities-are-endless/
51. https://oversight.house.gov/hearing/examining-sexual-harassment-gender-discrimination-u-s-department-agriculture/
52. http://news.nationalgeographic.com/news/2003/08/0808_030808_smokejumpers.html
53. Sheley, Chuck. "Is The Smokejumper Physical Fitness Test Eliminating Good Women Firefighters From Smokejumping?" Smokejumper Magazine posted: 2015-06-07 16:45:24.
54. Messenger, Lori. "Three Decades of Women in Smokejumping." Smokejumper Magazine, April, 2013. www.smokejumpers.com
55. https://www.eeoc.gov/laws/statutes/pregnancy.cfm

56. http://msmagazine.com/blog/2013/05/28/10-things-that-american-women-could-not-do-before-the-1970s/
57. http://www.huffingtonpost.com/2013/02/04/maternity-leave-paid-parental-leave-_n_2617284.html
58. Messenger, Lori. "Three Decades of Women in Smokejumping" Smokejumper Magazine, April, 2013. www.smokejumpers.com
59. Sinclair, Donna Lynn. "Caring for the Land Serving People: Creating a Multicultural Forest Service in the Civil Rights Era," PhD Dissertation, Portland State University, 2015.
60. https://www.fedscope.opm.gov
61. TimberLines Magazine April 1989, Volume XXVIII, 4. Region 6 U.S. Forest Service, Thirty-Year Club.
62. https://www.fedscope.opm.gov
63. https://www.fedscope.opm.gov
64. https://www.fedscope.opm.gov
65. http://www.military.com/benefits/veteran-benefits/veterans-employment-preference-points.html
66. https://www.fs.usda.gov/detail/alabama/news-events/?cid=STELPRD3804685
67. https://www.fedscope.opm.gov
68. https://www.fs.fed.us/sites/default/files/media/2015/16/Pathways_Applicant.pdf
69. Sharik, Terry L., and Stacy Frisk. "Reasons and Reservations for Enrolling in Forestry Degree Programs: A Survey of Undergraduate Students" *Western NAUFRP Meeting* (2008).
70. Sharik, Terry L., and Robert J Lilieholm. "Undergraduate Enrollment Trends in Natural Resources at NAUFRP Institutions: An Update" *Conference on University Education in Natural Resources (2010)*.

Bibliography

Albertson, Mary. "Progress of Women in the Forest Service," *Women in Natural Resources* 15 (September 1993): 4-5.

Anonymous. *Early Days in the Forest Service.* Missoula: U.S.D.A. Forest Service, Northern Region, 1944.

Anonymous. Rocky Mountain Retirees Association. *What Did We Get Ourselves Into?: Stories by Forest Service Wives.* Missoula: Rocky Mountain Retirees Association, 2000.

Bonta, Marcia Myers, ed. *American Women Afield: Writings by Pioneering Women Naturalists.* College Station: Texas A&M University Press. 1995, 248 p.

Bramwell, Lincoln. *Forest Management for All: State and Private Forestry in the U.S. Forest Service.* Durham: Forest History Society, 2013.

Dana, Samuel Trask. *Forest and Range Policy: Its Development in the United States.* New York: McGraw-Hill Book Company, 1956.

Duhse, Robert J. "The Saga of the Forest Rangers." *Elks Magazine,* July-August, 1986.

Enarson, Elaine Pitt. *Woods-Working Women: Sexual Integration in the U.S Forest Service.* Tuscaloosa: The University of Alabama Press, 1984.

Fisher, Carla. "You're Not Getting Rid of Me: Cultivating Space for Women in the U.S. Forest Service, 1950-1990." PhD dissertation, Purdue University, 2010. Pro Quest (UMI 3449744.)

Kaufman, Polly Welts. *National Parks and the Woman's Voice: A History.* Albuquerque: University of New Mexico Press, 1998.

Lewis, James G. *The Forest Service and the Greatest Good: A Centennial History.* Durham: Forest History Society, 2005.

_____. "The Applicant is No Gentleman: Women in the Forest Service," *Journal of Forestry* 103:5 (2005): 259-263.

Pendergrass, Lee F. "Dispelling Myths: Women's Contributions to the Forest Service in California," *Forest and Conservation History* 34 (1) (January 1990).

Reiner, Jacqueline S., ed. *An Interview with Geri Vanderveer Bergen*. Durham: Forest History Society, 2001.

_____. *An Interview with Wendy Milner Herrett*. Durham: Forest History Society, 2001.

_____. *An Interview with Leigh Beck*. Durham: Forest History Society, 2002.

_____. *An Interview with Clara Johnson*. Durham: Forest History Society, 2002.

Severance, Carol C., ed. *An Interview with Beverly C. Holmes*. Durham: Forest History Society, 2002.

Sharik, Terry L., and Robert J. Lilieholm. "Undergraduate Enrollment Trends in Natural Resources at NAUFRP Institutions: An Update" *Conference on University Education in Natural Resources* (2010).

Sharik, Terry L., and Stacy Frisk. "Reasons and Reservations for Enrolling in Forestry Degree Programs: A Survey of Undergraduate Students" *Western NAUFRP Meeting* (2008).

Steen, Harold K. *The Chiefs Remember: The Forest Service, 1952–2001*. Durham: Forest History Society, 2004.

_____. *The U.S Forest Service – A History*. Seattle: University of Washington Press, 1976.

West, Terry L. *Centennial Mini-Histories of the Forest Service*. Washington, D.C.: United States Department of Agriculture Forest Service, FS-518, 1992.

Williams, Gerald W. "Women in the Forest Service: Early Historical Accounts." Unpublished manuscript, U.S. History Collection, Forest History Society, Durham, NC.

Bibliography

Websites and Web Pages

"BIA Website." Indian Affairs. http://www.bia.gov/.

"BLM — The Bureau of Land Management." BLM — The Bureau of Land Management. http://www.blm.gov/wo/st/en.html.

"Fish and Wildlife Service Home Page." Fish and Wildlife Service Home Page. http://www.fws.gov/.

"Inventory of the Gene Bernardi Papers, 1971–1991." Inventory of the Gene Bernardi Papers, 1971 - 1991. http://www.foresthistory.org/ead/Bernardi_Gene.html.

Mayell, Hillary. "Women Smokejumpers: Fighting Fires, Stereotypes" (National Geographic News August, 2003) https://www.nationalgeographic.com/news.

Messenger, Lori. "Three Decades of Women in Smokejumping." (Smokejumper Magazine April, 2013.) www.smokejumpers.com.

National Smokejumper Training Guide. USFS 2008. Physical Conditioning Lesson Plan. 1.http://www.fs.fed.us/fire/aviation/av_library/sj_guide/05_physical_conditioning.pdf

Shulman, Deanne. "On Becoming a Smokejumper" (Smokejumper Magazine January, 2003) www.smokejumpers.com.

The U.S. Forest Service — An Overview http://www.fs.fed.us/documents/USFS_An_Overview_0106MJS.pdf

Traverse, Hannah. "SheSheShe" Camps: A Women's Alternative to the Civilian Conservation Corps. The Corps Network March 2013. http://corpsnetwork.org/shesheshe-camps-womens-alternative-civilian-conservation-corps. Further sources cited at the end of the article.

United States. National Park Service. "U.S. National Park Service About Us." National Parks Service. January 15, 2015. Accessed January 24, 2015. http://www.nps.gov/aboutus/index.htm.

"U.S. Forest Service." U.S. Forest Service. http://www.fs.fed.us/.

Index

accretion of duties, 371
Affirmative Action, 413, 423
agencies: other federal land management, 15–17. *See also* mission, agency
Alaska, Forest Service jobs in, 72, 139–141, 148, 150–154, 178, 180–186, 197, 199, 200, 202, 222, 224, 226–227, 240–243, 318, 351, 386, 388, 397
AmeriCorps National Civilian Community Corps (NCCC), 28, 410
archaeologist, 72, 100–104, 109–111, 137, 166
Audubon Society lawsuit: Forest Service jobs created as result of, 125

Baker, Sarah, engineer: biography of, 195–205; facilities engineer duties of, 199; non-traditional career choice, 201; Scotland in, 204–205; Siberia in, 203–204; sustainable operations coordinator and EMS program manager duties of, 197–198, 200, 205
Bard, Jane, silviculturist: biography of, 73–78; non-traditional career choice, 77; silviculture duties of, 76–77
Bernardi, Jean, 25, 438
Bernardi v. Butz, 25, 35. *See also* consent decree
Biaggi, Carmie, assistant dispatch center manager: biography of, 32–37; timber department duties of, 34
botanists, 28, 72, 80, 92, 94, 406
built environment image guide the, 207

Burnett, Kathy, wildlife biologist: biography of, 78–83

Canfield, Jodie, wildlife program manager: biography of, 113–116; non-traditional career choice, 116; wildlife biologist duties of, 115; wildlife program manager duties of, 115
career: life balance, 416, 436; support systems, 35, 68, 95–96, 146, 170–171, 212, 233, 236, 248, 325, 387, 405–406, 412, 419, 433
career and parenting, 86–87, 97, 170, 261–263, 325, 382, 388, 404, 412, 418–419
career ladders, 263, 423, 427
Champion, Cindy, smokejumper: biography of, 217–226; non-traditional career choice, 222; smokejumping duties or tasks of, 4, 221–222; smokejumping physical training for, 221; physical standards for, 223. *See also* smokejumpers, smokejumping
chief of the Forest Service, 12, 19, 22, 24, 29, 193, 395; chief's staff role, 311, 399
civil rights movement, 3, 25, 27, 30,
Civilian Conservation Corps (CCC), 24; as model for youth and conservation programs, 26
clerical: clerk, 3, 23, 30, 80, 108, 131, 180, 189, 314, 392, 402, 403, 425
Connolly, Stephanie, soil scientist: biography of, 116–123; duties of, 119–121; non-traditional career choice, 121

445

consent decree California, 3, 25-26, 30, 35, 41, 81, 108, 111, 325, 403, 413
conservation, vii, x, 3, 6, 8, 24, 26-29, 148, 197, 207, 209, 215, 261, 268, 319, 321, 332, 333, 360, 392, 434
Craig, Garret, deputy district ranger and former disaster program specialist, International Programs: biography of, 226-245; Alaska, jumping in, 227, 240-242; contract fire crew on, 229-230; DASP duties in, 245; hotshot crew, challenges of being on, 231-234; non-traditional choice, 231; rappeller duties of, 235; smokejumping challenges, 237-241; smokejumping physical maladies of, 242, 244; smokejumper rookie training, 236
cross-training, 73, 85, 412
Cueva, Susan, hydrology technician: biography of, 37-51: backcountry patrol duties of, 44-45; brush disposal duties of, 42; fire fighter duties of, 41-42; hydrology technician duties of, 49-50; recreation technician duties of, 47; trail crew duties of, 45; wild river patrol duties of, 46

Department of Agriculture, 17, 22, 55, 311, 414
Department of the Interior, 17, 22, 55, 133, 308
detail, job, 35, 69, 75, 85, 86, 102, 111, 133, 151, 191, 198, 200, 205, 209, 215, 217, 242, 284, 315, 318, 321, 397, 409, 420
discrimination. 3, 8, 25, 32, 68, 96, 162, 202, 251, 269, 276, 340, 351, 387, 392, 414, 419
district rangers, 22, 24, 29, 72, 113, 217, 304, 316, 345, 350, 371, 379, 386, 392, 409, 410, 422

diversity, workplace, 161, 179, 193, 320, 393, 394, 413, 420, 423, 425, 433, 434, 436
dual careers, x, 208, 380, 406, 412, 426-427

Earth First, 367
ecologists, 134, 141, 356
Endangered Species Act (ESA) of 1973, 27, 28
engineers: duties of, 12, 18, 72, 157, 166, 194-195, 197, 199, 200, 207-208, 256, 260, 367, 368, 376
entomologists, 72, 213-214, 284, 289-297, 422
environmental management system (EMS), 197-198
environmental movement, 27, 30, 412
Erickson, Mary, forest supervisor: biography of, 377-382; analyst and economist job advantages of, 379; forest supervisor duties of, 380; non-traditional career choice, 380
Estill, Elizabeth: as first female regional forester, 29
expanded dispatch, 69

Fairbanks, Kristine, law enforcement officer: biography of, 245-251; dangers of law enforcement work, 250, 251; K-9 law enforcement, 248-250; non-traditional career choice, 248; memorial to, 249
Feigley, Rachel, district ranger: biography of, 312-318; Alaska detail experience, 318; biologist duties of, 314-315; district ranger transition to, 316; non-traditional career choice, 314; wilderness guard seasonal positions, 317-318; YCC skills learned in, 313
Fenner, Patti, forest noxious weed program manager: biography of, 123-129; duties of, 128-129; range

conservationist duties of, 125–127; recreation technician duties of, 124–125

fire fighting, 16, 17, 22, 26, 29, 31, 34, 42, 64, 65, 78, 85, 216, 217, 220, 223, 229, 230, 288, 231, 338, 356, 358, 408, 416, 424

first female, first woman, 12, 23–24, 29, 61, 172, 193, 269, 289, 290, 350, 387, 388, 395, 418, 432

fisheries biologists, 28, 72, 94, 106, 392

Forest Reserve Act, 19

Forest Service, organization of, 18

forest supervisor, 29, 113, 305, 311, 377, 380, 382, 386, 393, 394, 398, 422

foresters, 27, 28, 75, 80, 83, 85, 94, 131, 132, 288, 290, 303, 351, 364, 385, 392, 396, 397, 421–422, 432

Fuller, Molly, district ranger: biography of, 318–334; consent decree on, 325; ecosystem managers, team bond, 331; experiences remembered, 331–332; harassment experienced, 324; resource officer assistant skills learned as, 322

general schedule (GS) pay plan, 25, 73, 269, 423, 427

Geospatial Service and Technology Center (GSTC), 194

Girl Scouts: as influence in outdoors interest, 23, 168, 177, 384

Grantham, Patty, forest supervisor: biography of, 382–390; district ranger duties of, 386; district staff officer duties of, 386; fire crew boss first female on, 388; forester duties of, 385; forest supervisor advantages of being, 386; land adjustment duties, 385–386; non-traditional career choice, 385; parenting challenges, 388

greatest good, xiii, 19, 24, 415, 436

gridding, in firefighting, 230

Hazlitt, Cheryl, interpretive planner: biography of, 251–266; enterprise team as, 262; international assignments, 216, 257, 263; interpretation, on working in, 254; interpretive planner characteristics of, 256; interpretive planner duties of, 255; Nepal, travel to, 253, 254, 263, 265; non-traditional career choice, 259; work and parenting challenges on, 261–262, 263

Herger-Feinstein Quincy Library Group Forest Recovery Act (HFQLG), 157, 330

Herrett, Wendy: as first female district ranger, 29

Hoie, Sonja, recreation program manager: biography of, 167–177; lands and special uses administrator duties of, 173; NPS backcountry ranger duties of, 172–173; non-traditional career choice, 169; recreation program manager duties of 175–176; recreation staff officer duties as, 173; supervisory natural resources specialist duties of, 174–175

House Committee on Oversight and Government Reform 2016 hearings on sexual harassment, 414

Humphrey, Beth, district ranger: biography of, 334–348; district ranger duties of 345–346; forest wildlife biologist duties of, 344–345; harassment experience of, 338–340; non-traditional career choice, 338; range conservationist duties of, 341–342; range technician duties of, 341; wildlife biologist duties of, 342; wildlife, fisheries and rare plants staff duties of, 342–343; wildlife staff for Pawnee National Grasslands duties of, 343–344

Hydrologists, 28, 72, 93, 94, 134, 154, 158, 215

interagency hotshot crew (IHC), 65
interdisciplinary planning teams, interdisciplinary teams (IDTs), 28, 72, 84, 85, 102, 103, 132, 177, 342, 356, 406
International Programs, International Forestry, 7, 18, 213, 215–217, 226, 242, 244–245, 257, 263, 360, 393, 394

Job Corps, 26, 213, 298, 306–307. *See also* War on Poverty. *See also* youth and conservation programs
Job Corps Civilian Conservation Centers (JCCCCs) program emphasis, 26
Jones-Crabtree, Anna, engineer: biography of, 205–212; duties of, 207–209; Greater Yellowstone Coordinating Committee's Sustainable Operations Subcommittee as founding co-chair of, 209; non-traditional career choice, 206

Kimbell, Abigail (Gail), chief emeritus: biography of, 395–400; positions held, 396–397; career vs. family challenges on, 400; chief's position challenges of on, 398–399; committee on forestry meeting, Italy on, 399; discrimination on, 400; non-traditional career choice, 399; Washington Office personnel on, 398
Kluwe, Joan, recreation planner: biography of, 177–186; non-traditional career choice, 179; recreation planner duties of, 183; wilderness ranger duties of, 180
Krivacek, Janet, district ranger: biography of, 348–352; discrimination experience of, 351; district ranger on being a, 352; non-traditional career choice, 350; silviculture certification process on, 351

Land Management Workforce Flexibility Act, 31, 423, 424
land use planners, 18, 28, 72, 85
landscape architects, 28, 29, 166, 207, 256, 260
Lane, Cornelia (Connie), recreation program manager: biography of, 186–193; rewards of the job on, 189; non-traditional career choice, 191
Larson, Geri: as first female forest supervisor, 29
law enforcement: agency, 18, 29, 31, 172, 173, 217, 245, 247, 279, 280, 305, 416, 428; K-9, 248, 249, 280; level II classification, 48, 342
Leach, Anita, forester: biography of, 83–87; discrimination experienced, 86; forester duties of, 85; non-traditional career choice, 86; reforestation duties of, 85; planner duties of, 85
line officers, line positions, 72, 152, 163, 304, 311–312, 316, 356

male-dominated field: challenges for women working in, 41, 61, 68, 107, 111, 121, 144, 145, 169, 281, 296, 338, 380, 412, 413, 414, 432
"March for Our Lives," 415
masculine behaviors as coping mechanism, 416
"Me Too" movement, 415
militia, fire, 34, 69, 356, 408
Millar, Connie, senior scientist: biography of, 266–278; challenges, institutional on, 276; field season duties of, 272–275, geographical focus of, 275; forest research branch and NFS relationship, 270; imposter phenomenon, 272; non-traditional career choice, 268; scientist, self-image as, 270–272; senior technical scientist designation, 269; winter duties of, 275–276

448

Index

Mills, Lynda Perry, wildlife biologist: biography of, 87-100; minority in wildlife classes, as a, 90; non-traditional career choice, challenges of, 90-91, 94-96; red-cockaded woodpecker recovery, her role in, 93; wildlife biologist, duties of, 92-94

mission: Forest Service, 15, 18, 112, 129, 148, 193, 433, 436; Forest Service vs. NPS confusion, 17; other federal land management agencies; BIA, 16; BLM, 15-16; NPS, 16-17; USFWS, 16. *See also* agencies, other federal land management

motto, Forest Service: Caring for the Land and Serving People, 15, 18

Muchowski, Mary, biological science technician: biography of, 51-63; non-traditional career choice, 61; wildlife refuge duties of, 56, 58; wildlife technical duties of, 54-55, 57-58

Multiple-Use Sustained-Yield Act of 1960: shift in Forest Service management priorities, 18, 27

National Environmental Policy Act (NEPA) of 1970, 27, 28, 50, 54, 75, 85, 102, 159, 160, 186, 316, 342, 356

National Forest System: organization of, 7, 18, 19, 20, 213, 270, 393, 394

natural resource careers, traditional vs. other disciplines transition to, 421-422

non-competitive job eligibility, 75

non-traditional career choice, 7, 15; *See also under* individual biographies

Office of Personnel Management, 14

"ologists," 28, 131, 132; specialists, 28, 72, 73, 113, 132, 198, 314, 320, 356

pack test, fire fighting, 78, 223, 236, 358, 359

Page, Roberta, law enforcement officer: biography of, 279-284; discrimination experienced, 282; duties of, 280-281; K-9 officer as, 280, 284; non-traditional career choice, 281-282

Parker, Wendy, archaeologist: archaeology duties of, 101-103; biography of, 100-104; heritage helpdesk duties of 102; heritage data steward duties of, 102; non-traditional career choice, 103

Parrie, Traute, district ranger: biography of, 352-362; Africa in, 360-362; feminist movement influence of, 355; fire fighting hardship and rewards of, 358-359; fire militia roles, 356, 358

Pathways program, 122, 123, 426

Pinchot, Gifford: first chief, vii, xiii, 19, 22, 29

Pollock, Nadine, ecosystems staff officer: biography of, 129-134; details to other jobs, 133; ecosystems staff officer duties of, 132; natural resources staff officer duties of, 133

Pregnancy Discrimination Act of 1978, 419

professional positions: requirements for, 30, 80, 426

Ragenovich, Iral, regional entomologist state and private forestry: biography of, 284-298; duties of, 291, 292; 293-294; NFS lands, 294; non-traditional career choice and challenges of 287, 294-297; other federal lands, private lands, service to, 294; Russia in, 294

ranger districts, 19, 56, 72, 73, 113, 217

recreation managers, recreation management specialists: duties of, 28, 165, 166; recreation technicians duties of, 165
regional forester, 311, 422
regional offices, role of, 19, 113
Research and Development, 7, 18
Roberts, Cindy, wildlife and aquatic biologist: biography of, 104–108; non-traditional career choice, challenges of, 107–108; work duties of, 106–107

Senior Executive Service Candidate Development Program (SESCDP), 307
sexism, sexist, 90, 95, 388, 417
sexual harassment: 61, 108, 170, 268, 276, 414, 417; culture of, 413, 414, 415
SheSheShe camps, 24
Sholly, Deb, minerals management staff: biography of, 134–147; seasonal jobs of, 134–135, 137–139, 141; male-dominated jobs on, 144; minerals and special uses duties of, in Alaska, 141; non-traditional career, challenges of, 144–146
smokejumpers, smokejumping, physical requirements of, 217, 223, 417; women as normal in, 417, 418
Social movements; of 1960s and 1970s, 27
soil scientists, 28, 72, 93, 94, 116
Stanley, Barbara, Forest and Regional Energy Coordinator and Program Manager: biography of, 147–154; Forest Service technician duties of, 149–150; non-traditional career choice, 153; progress of career and duties of, 150–153; recreation research technician, duties of, 149; temporary jobs in Alaska, duties of, 150

Student Conservation Association (SCA), 24, 25, 175
State and Private Forestry, 7, 18, 213, 290, 393, 394
supervisor's offices: organizational level of, place in Forest Service organization; role of, 113, 194, 213
sustainable operations, 197, 198, 200, 205, 208, 209, 211–212
sustainable operations collective, 12, 208, 209

technicians, disciplines and general duties of, 30, 31–32, 165; physical demands of, 32; qualification requirements for, 30
technology and development centers, 194
Terrell, Tina, national director of the job corps: biography of, 298–310; African-American Program Manager experience of, 302; district ranger responsibilities as, 304; forest supervisor challenges as, 305–306, 308–309; inventory forester duties of, 303; job corps duties and accomplishments in, 306–307; legislative affairs specialist accomplishments as, 305, 307–308; liaison officer accomplishments as, 303–304; non-traditional career choice, 307; small sales officer duties of, 302
Tibbetts, Deborah, archaeologist: biography of, 109–112; non-traditional career choice, 111
Tonto National Forest: friends of, 129
Turner, Lauren, district ranger: biography of, 401–411; biological technician duties of, 404; details to other jobs, 409; district ranger duties of, 409–410; ecosystem manager duties of, 409; federal women's program manager

collateral duties of, 408; non-traditional career choice, 403; consent decree opportunities on, 403; wildlife biologist duties of, 406–408

Type I fire crew: hotshot crew, 65–66, 68

veteran's preference, 31, 423, 424

Volunteers in Service to America (VISTA), 28, 410. *See also* AmeriCorps

Vonderheit, Erin, fire dispatcher: biography of, 63–71; initial attack dispatcher duties of 69–70

War on Poverty, 26. *See also* Job Corps

Weldon, Leslie, deputy chief for the National Forest System: biography of, 390–395; discrimination on, 392; non-traditional career choice, 391; career progression, variety of positions and responsibilities in, 392–394; work-life balance on, 394–395; YCC experience in, 391–392

Whitsett, Kelly, Forest Hydrologist and Cave and Karst Program Manager: biography of, 154–164; hydrologist duties of, 157–158; hydrologist and cave and karst program manager duties of, 159–160; non-traditional career choice, 160–161

Wilderness Act of 1964, 27

wildlife biologists, 28, 56, 72, 80, 94, 215

wives, vii, 13, 23, 32

Wooding, Ruth, last Sula district ranger: biography of, 362–376; accessibility coordinator duties of, 368–369; AutoCAD technician duties of, 368; district ranger duties of, 371–374; Earth First monkey wrenchers strategy against, 367–368; firefighting duties of, 367; lands and easement administrator duties of, 370–371; non-traditional career choice, 374–375; reforestation and silviculture duties of, 366

Young Adult Conservation Corps (YACC), 26, 27, 280

youth and conservation programs, 26. *See also* Job Corps

Youth Conservation Corps (YCC), 26, 27, 80, 313, 391